STUDY GUIDE
EMILY J. KEATON

INTERMEDIATE ALGEBRA

SECOND EDITION

K. ELAYN MARTIN-GAY

PRENTICE HALL
Upper Saddle River, NJ 07458

Acquisitions Editor: *Melissa Acuña/Ann Marie Jones*
Production Editor: *Kimberly Dellas*
Special Projects Manager: *Barbara A. Murray*
Supplements Editor: *April Thrower*
Art Director: *Amy Rosen*
Production Coordinator: *Alan Fischer*

© 1997 by **PRENTICE-HALL, INC.**
Simon & Schuster/A Viacom Company
Upper Saddle River, NJ 07458

Printed in the United States of America

10 9 8 7 6 5 4 3 2 1

ISBN 0-13-258088-8

Prentice-Hall International (UK) Limited, *London*
Prentice-Hall of Australia Pty. Limited, *Sydney*
Prentice-Hall Canada, Inc., *Toronto*
Prentice-Hall Hispanoamericana, S.A., *Mexico*
Prentice-Hall of India Private Limited, *New Delhi*
Prentice-Hall of Japan, Inc., *Tokyo*
Simon & Schuster Asia Pte. Ltd., *Singapore*
Editora Prentice-Hall do Brasil, Ltda., *Rio de Janeiro*

Preface

This Study Guide accompanies *Intermediate Algebra,* Second Edition, by K. Elayn Martin-Gay. It provides you with additional resources to help you be successful in your mathematics course.

This Study Guide begins with a list of study skills and note-taking suggestions, as well as guidelines for working in groups, titled "Study for Success." These study hints follow the Table of Contents.

For each section in the text, there is a corresponding section in the Study Guide giving additional worked-out examples and a set of additional practice exercises. The examples and exercises match the style of the text. Exercises that require the use of a grapher, such as a graphing calculator or computer graphing software, are marked with the symbol $\mathbf{C}$. You will find answers to all of the exercises, along with selected exercise solutions, in the back of the Study Guide.

Each chapter of this Study Guide concludes with a list of Hints and Warnings and a Practice Chapter Test. The Hints and Warnings sections summarize mathematical methods and problem-solving procedures, as well as warn against common errors. Each Practice Chapter Test is similar to the chapter tests given in the text. In addition, two different 100-question Practice Final Exams are included; one exam consists of open-ended questions, and the other exam consists of multiple-choice questions. The answers to all tests and exams are given in the back of the Study Guide.

I would like to thank senior editor Ann Marie Jones for the opportunity to work on this project, editorial assistant April Thrower for facilitating this project, and James Sellers at Cedarville College for his excellent accuracy review. I would also like to thank my husband, Mark, for his unfailing understanding and daily support throughout this project. I am grateful to my parents and sister, Amy, for their long-distance encouragement.

Emily J. Keaton

Table of Contents

Study for Success

Consider the following tips for becoming more successful in your mathematics course.

- Attend each class. If you must miss a class, borrow notes from another student and make a copy. Call and get the homework assignment so that you do not get behind. Consider viewing the videotapes that accompany this textbook if they are available at your school's library or media resource center.

- In your mathematics classroom or lab, be sure you have a good view of the board or overhead projector screen and that you can hear your instructor or lab leader.

- Come to class prepared. Use a notebook just for this class, preferably consisting of a section for notes and a section for homework. Bring the appropriate writing utensils.

- You may be forced to take notes quickly during class. Set up your own abbreviation system. Take notes as thoroughly as possible. After class you can rewrite your notes more legibly and fill in any gaps.

- It is important to keep good notes in your mathematics class because these will become yet another resource when you do your homework, study for exams, and prepare for later courses. If you can't remember how to solve a problem when doing your homework, check your notes to see how your instructor suggested to do the problem. You can also refer to the examples in the textbook or this Study Guide.

- Be sure you understand what has been assigned for homework. You can always double-check the assignment with another student in the class. Do your homework after class as soon as possible. This way the information is still fresh in your mind.

- You may find it helpful to find one or more study partners. Consider working together after class to review the material covered during class and to be available as resources to one another while completing homework assignments. You may also find that explaining concepts to a study partner helps to clarify the concepts in your own mind. When preparing for tests and exams, you and your study partners may find it helpful to quiz each other on major concepts.

- Practice is what improves your skills in every area, especially mathematics. Practice is what will make you become comfortable with problems. If there are certain assigned problems that you are having difficulty with, do more of this type until you can do them easily. Remember that this Study Guide provides you with lots of additional problems for practice.

- As you do homework problems and read the textbook, write down your questions. Don't rely on your memory, you may have forgotten an important part of your question by the time you reach your class.

- As you read your textbook, don't just read the examples. The entire section is important in your understanding of the material. Don't forget that this Study Guide provides additional examples if you require further clarification. The numerous additional exercises in this Study Guide will also give you extra practice on any topics for which you need it.

- Don't be afraid to ask questions. If something is not clear, ask for further explanation. Usually there are other students with the same questions.

- Be sure to hand in assignments on time. Don't lose points needlessly.

- Take advantage of all extra-credit opportunities if they exist.

- When doing homework or taking a test, give yourself plenty of paper. Show every step of a problem in an organized manner. This will help reduce the number of careless mistakes.

- When you are studying for a test or exam, you may find the practice chapter tests in this Study Guide a helpful resource of review and self-assessment. There are also two different practice final exams included in this Study Guide.

- It is important to study your mathematics as many days as you can possibly fit into your schedule. Even if it is just for 10 or 15 minutes, this is beneficial. Do not wait until just before a test to do a crash study session. The material should be understood and learned, not memorized. The exercises in this Study Guide provide you with additional resources for practicing your skills and keeping them current. Most likely you will take another mathematics course after this one. You are gaining tools in this class that you will use in the next class.

- Many times students can do their homework assignments but then "freeze up" on a test. Their minds go blank, they panic, and, hence, they do not do well on their test. If this happens to you, it can be overcome by practicing taking tests. When studying for a test, make up an actual test for yourself or use one if the practice tests in this Study Guide if appropriate. Then find a quiet spot and pretend you are in class. Take this test just as if you were in class. The more you practice taking tests, the more comfortable you will be in class during the actual event.

Consider the following guidelines for working cooperatively in groups:

- Be sure to have each group member introduce himself or herself to the rest of the group.
- Agree on what your group must do and how you will get it done.
- Be courteous and listen carefully to other group members.
- Create an atmosphere that allows group members to be comfortable in asking for help when needed.
- Remember that not everyone works at the same pace. Be patient. Offer your help. Receive help graciously.
- Keep in mind that all contributions made by group members are valuable.
- Ask for your instructor's help only when all other sources of assistance have been exhausted.
- If you finish early, double-check your work. As appropriate, reflect on your work. Can you make a general rule about the solution or describe a real-world use for what you learned?

CHAPTER 1

Real Numbers and Algebraic Expressions

1.1 Algebraic Expressions and Sets of Numbers

EXAMPLE 1 *Objective 1: Identify and evaluate algebraic expressions.*

A triangular window has a height h of 50 centimeters and a base b of 42 centimeters. Use the algebraic expression $\frac{1}{2}bh$ to find the area of the window.

Solution: Replace b with 42 and h with 50 in the algebraic expression $\frac{1}{2}bh$.

$$\frac{1}{2}bh = \frac{1}{2} \cdot 42 \cdot 50 = 1050 \text{ square centimeters.}$$

EXAMPLE 2 *Objective 2: Identify natural numbers, whole numbers, integers, rational and irrational real numbers.*

Determine whether the following statements are true or false. Explain your reasoning.
  a. 0 is a natural number.
  b. $\{-2, 0, 2, 4\} \subseteq \{x | x \text{ is a whole number}\}$
  c. Every integer is a rational number.
  d. $0.25 \in \{x | x \text{ is a rational number}\}$
  e. Some integers are irrational numbers.

Solution: a. False. The number 0 is not included in the natural numbers; however, it is included in the whole numbers.
  b. False. Because the number -2 is not included in the whole numbers, the set $\{-2, 0, 2, 4\}$ is not a subset of the whole numbers.
  c. True. Every integer can be represented as a ratio of two integers.
  d. True. Because the number 0.25 is equal to the ratio of the integers 1 and 4, 0.25 is an element of the rational numbers.
  e. False. Because integers are rational numbers and irrational numbers are defined as any real number that is not a rational number, no integers are irrational numbers.

EXAMPLE 3 *Objective 3: Write phrases as algebraic expressions.*

Translate each phrase to an algebraic expression. Use the variable y to represent each unknown number.
  a. A number divided by thirteen
  b. Five less than twice a number
  c. The total of 10, 6, and one-third of a number
  d. The product of a number and 125
  e. A number subtracted from four

Solution: a. "A number"   "divided by" "thirteen"

↓	↓	↓
y	$\div$	13

$\rightarrow\ y \div 13$

b. "Twice a number" translates to $2y$, and "less than" indicates subtraction. Therefore, "five less than twice a number" translates to $2y - 5$.

c. "Total" indicates addition, and "one-third of a number" translates to $\frac{1}{3}y$. Therefore, "the total of 10, 6, and one-third of a number" translates to $10 + 6 + \frac{1}{3}y$.

d. $125y$

e. $4 - y$

EXERCISES

Exercises 1–8. Find the value of each algebraic expression at the given replacement value. See Example 1.

1. $\frac{1}{2}x$ when $x = 4$ **2.** $3r$ when $r = 5$ **3.** $-14z$ when $z = 2$ **4.** $\frac{3}{5}y$ when $y = -5$

5. xy when $x = 2$ and $y = 12$ **6.** $\frac{1}{3}r - s$ when $r = 9$ and $s = 2$

7. $2ab - 3$ when $a = 4$ and $b = 5$ **8.** $\frac{x}{y} + 5z$ when $x = 10$, $y = 2$, and $z = -1$

9. Use the algebraic expression bh to find the area of a parallelogram with a base b of 15 inches and height h of 32 inches.

10. Use the algebraic expression $\frac{1}{2}bh$ to find the area of a triangle with a base b of 2.5 centimeters and a height h of 3.4 centimeters.

Exercises 11–14. List the elements in each set.

11. $\{x | x$ is a whole number between -2 and $3\}$

12. $\{x | x$ is a natural number less than $-1\}$

13. $\{x | x$ is an integer between 0 and 6$\}$

14. $\{x | x$ is an even positive integer$\}$

Exercises 15–20. List the elements of the set $\{2, \frac{17}{4}, \pi, -\frac{1}{2}, \sqrt{49}, 0, -\sqrt{49}, \frac{12}{3}, \sqrt{50}\}$ that are also elements of the given set.

15. Whole numbers   **16.** Rational numbers   **17.** Real numbers   **18.** Integers

19. Natural numbers

20. Irrational numbers

Exercises 21–26. Write each phrase as an algebraic expression. Use the variable z to represent each unknown number. See Example 3.

21. Twice a number

22. Three less than one-half of a number

23. The quotient of a number and 11

24. Ten less than a number

25. The difference of a number and four times a number

26. The sum of one hundred times a number added to 6

27. One serving of a recipe contains 360 calories and 6 grams of fat. The algebraic expression $9f \div c$, where f represents number of fat grams per serving and c represents number of calories per serving, gives the percent of calories from fat. Find the percent of calories from fat for this recipe.

28. It takes 39,090 gallons of water to manufacture a new car. The algebraic expression $39,090x$ gives the amount of water used to manufacture x cars. How much water is used to manufacture 25 cars? (data from American Water Works Association)

29. The bar graph below shows the total annual revenue for the leading businesses in the transportation equipment industry for 1994. (Note that dollar amounts are in millions.) (data from *Fortune* Magazine)
   a.   Which company had the most revenue?
   b.   Which company had the least revenue?
   c.   What is difference between the lowest and highest revenues?

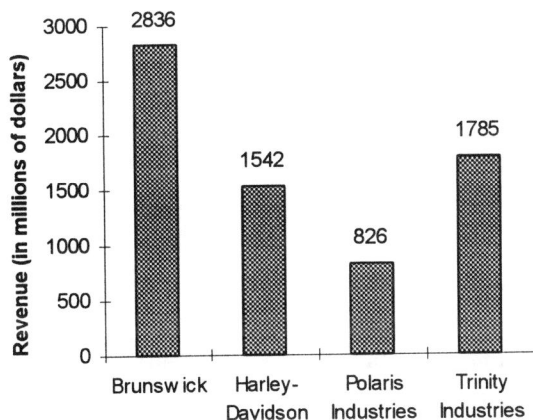

1.2   Properties of Real Numbers

EXAMPLE 1   *Objective 1: Use operation and order symbols to write mathematical sentences.*

Write each sentence using mathematical symbols. Use the variable t to represent any unknown number.

    a. One-half of a number is less than or equal to 13.
    b. The total of a number and 2 yields three times z.
    c. The quotient of 6 and a number is greater than 17.
    d. The product of a number and -3 is not equal to the sum of x and 5.

Solution: **a.** $\dfrac{1}{2}t \le 13$

             b. $t + 2 = 3z$

             c. $\dfrac{6}{t} > 17$

             d. $-3t \ne x + 5$

EXAMPLE 2   *Objective 2: Identify identity numbers and inverses.*

Write the additive inverse and the multiplicative inverse of each.

    a. -2 **b.** $\dfrac{3}{5}$ **c.** π **d.** $-\dfrac{12}{7}$

Solution: **a.**   The additive inverse of -2 is 2. The multiplicative inverse of -2 is $-\dfrac{1}{2}$.

             b.   The additive inverse of $\dfrac{3}{5}$ is $-\dfrac{3}{5}$. The multiplicative inverse of $\dfrac{3}{5}$ is $\dfrac{5}{3}$.

             c.   The additive inverse of π is $-\pi$. The multiplicative inverse of π is $\dfrac{1}{\pi}$.

             d.   The additive inverse of $-\dfrac{12}{7}$ is $\dfrac{12}{7}$. The multiplicative inverse of $-\dfrac{12}{7}$ is $-\dfrac{7}{12}$.

EXAMPLE 3   *Objective 3: Identify and use the commutative, associative, and distributive properties.*

State the property (commutative, associative, or distributive) of addition or multiplication that is illustrated in each statement.

    a. $(3 + x) + 5 = 3 + (x + 5)$
    b. $4 \cdot 6 = 6 \cdot 4$
    c. $-4(a + 2b) = -4a - 8b$

Solution: **a.**   The associative property of addition
             b.   The commutative property of multiplication
             c.   The distributive property

EXERCISES

Exercises 1–8. Write each statement using mathematical symbols. Use the variable z to represent any unknown number. See Example 1.

1. The quotient of -7 and y is 6.   **2.** Nine times the difference of 5 and a number amounts to -3.

3. The sum of 9 and a number is 18.   **4.** Three times the sum of b and 7 is 56.

5. The quotient of t and 3 is less than 11.   **6.** The difference of 4 and a number is not equal to 15.

7. Twice the sum of q and 8 is greater than or equal to the opposite of 12.

8. Five times the difference of a number and 13 is less than or equal to the reciprocal of $\dfrac{1}{2}$.

Exercises 9–16. Write (a) the additive inverse and (b) the multiplicative inverse of each number if they exist. See Example 2.

9. $-\dfrac{3}{13}$ **10.** 0   **11.** -4 **12.** 17   **13.** $\dfrac{5}{3}$ **14.** $\dfrac{1}{9}$ **15.** $-\dfrac{21}{2}$ **16.** -9

Exercises 17–23. State the property (commutative, associative, or distributive) of addition or multiplication that is illustrated in each statement. See Example 3.

17. $5 + 2x = 2x + 5$ **18.** $3a + (2b + 4) = (3a + 2b) + 4$ **19.** $(4 \cdot 5)t = 4(5t)$ **20.** $-7z = z(-7)$

21. $3(r + 2t) = 3r + 6t$ **22.** $(a + 3b) + 4c = a + (3b + 4c)$ **23.** $-2x - 2y = -2(x + y)$

Exercises 24–27. Use a commutative property to write an equivalent expression.

24. $-3x$ **25.** $p + 4q$ **26.** yz **27.** $3r + s$

Exercises 28–31. Use an associative property to write an equivalent expression.

28. $(5r + s) + 3t$ **29.** $2w + (3v + 17u)$ **30.** $3(ab)$ **31.** $(-5.3x)y$

Exercises 32–35. Use the distributive property to write an equivalent expression.

32. $2(1 + r)$ **33.** $z(7 + 6w)$ **34.** $3(x + y + z)$ **35.** $7(2a + 2b + 2c)$

1.3 Operations on Real Numbers

EXAMPLE 1 *Objective 1: Find the absolute value of a number.*

Find each absolute value.
a. $|7|$ b. $|-9|$ c. $-|-3|$

Solution: a. $|7| = 7$ because 7 is located 7 units from 0 on the number line.

b. $|-9| = 9$ because -9 is 9 units from 0 on the number line.

c. $-|-3| = -3$. The negative sign outside of the absolute value bars means to take the opposite of the absolute value of -3.

EXAMPLE 2 *Objective 2: Add and subtract real numbers.*

Find each sum or difference.
a. $-5+6$ b. $7+(-15)$ c. $-2-9$ d. $3-(-16)$

Solution: a. $-5+6=1$ b. $7+(-15)=-8$

c. $-2-9=-2+(-9)=-11$ d. $3-(-16)=3+16=19$

EXAMPLE 3 *Objective 3: Multiply and divide real numbers.*

Find each product or quotient.

a. $(-4)(-5)$ b. $3\left(-\dfrac{2}{9}\right)$ c. $\dfrac{-48}{-12}$ d. $\dfrac{-81}{3}$ e. $\dfrac{5}{6} \div (-5)$

Solution: a. Because the signs of the two numbers are the same, the product is positive. Thus $(-4)(-5) = +20$ or 20.

b. Because the signs of the two numbers are different, the product is negative. Thus

$$3\left(-\frac{2}{9}\right) = -\frac{6}{9} = -\frac{2}{3}.$$

c. Because the signs of the two numbers are the same, the quotient is positive and $\dfrac{-48}{-12} = 4$.

d. Because the signs of the two numbers are different, the quotient is negative and $\dfrac{-81}{3} = -27$.

e. $\dfrac{5}{6} \div (-5) = \dfrac{5}{6} \cdot -\dfrac{1}{5} = -\dfrac{1}{6}$

EXAMPLE 4 *Objective 4: Simplify expression containing exponents.*

Simplify each expression.

a. 4^3 b. $(-2)^4$ c. $(-3)^3$ d. $\left(\dfrac{2}{3}\right)^3$

Solution: a. $4^3 = 4 \cdot 4 \cdot 4 = 64$ b. $(-2)^4 = (-2)(-2)(-2)(-2) = 16$

c. $(-3)^3 = (-3)(-3)(-3) = -27$ d. $\left(\dfrac{2}{3}\right)^3 = \left(\dfrac{2}{3}\right)\left(\dfrac{2}{3}\right)\left(\dfrac{2}{3}\right) = \dfrac{8}{27}$

EXAMPLE 5   *Objective 5: Find roots of numbers.*

Find the following roots.

   a. $\sqrt{36}$ **b.** $\sqrt[3]{64}$ **c.** $\sqrt{49}$ **d.** $\sqrt[4]{16}$

Solution: **a.** $\sqrt{36} = 6$ because 6 is positive and $6^2 = 36$.

   b. $\sqrt[3]{64} = 4$ because $4^3 = 64$.

   c. $\sqrt{49} = 7$ because 7 is positive and $7^2 = 49$.

   d. $\sqrt[4]{16} = 2$ because 2 is positive and $2^4 = 16$.

EXERCISES

Exercises 1–4. Find each absolute value. See Example 1.

1. $|-16|$ **2.** $|55|$ **3** $-|-2|$ **4.** $-|7|$

Exercises 5–8. Find each sum or difference. See Example 2.

5. $-4 + (-3)$ **6.** $-17 + 7$ **7.** $5 - 12$ **8.** $-12 - (-37)$

Exercises 9–12. Find each product or quotient. See Example 3.

9. $-4(12)$ **10.** $(-3)(-10)$ **11.** $-14 \div -7$ **12.** $\dfrac{-18}{3}$

Exercises 13–16. Simplify each expression. See Example 4.

13. $(-4)^2$ **14.** -2^4 **15.** 5^3 **16.** $\left(\dfrac{1}{3}\right)^3$

Exercises 17–20. Simplify each expression. See Example 5.

17. $\sqrt{121}$ **18.** $\sqrt{25}$ **19.** $\sqrt[4]{81}$ **20.** $\sqrt{\dfrac{4}{25}}$

21. At the beginning of the month, the balance of your checking account is – 27 dollars. After two deposits your balance is 65 dollars. What was the total amount that you deposited?

22. At the beginning of the month, the balance of your checking account is 215 dollars. What is your balance after writing a check for 175 dollars?

23. On one day in January, the high temperature in Telluride, Colorado, was 36° Fahrenheit and the low temperature was − 7° Fahrenheit. Find the difference in temperatures.

24. The coldest temperature ever recorded in the U.S. was recorded as − 80° Fahrenheit in Prospect Creek, Alaska, in 1971. The world's coldest temperature of − 128.6° Fahrenheit was recorded at the Soviet Antarctica station Vostok in 1983. What is the difference between these two record low temperatures? (data from National Climatic Data Center)

1.4   Order of Operations and Algebraic Expressions

EXAMPLE 1   *Objective 1: Use the order of operations.*

Simplify.

$$\frac{(3-7)^2 - 2^2 \cdot 3}{-5 - (-3)}$$

Solution:

$$\frac{(3-7)^2 - 2^2 \cdot 3}{-5 - (-3)} = \frac{(-4)^2 - 2^2 \cdot 3}{-5 - (-3)} \qquad \text{Simplify inside grouping symbols first.}$$

$$= \frac{16 - 4 \cdot 3}{-5 - (-3)} \qquad \text{Raise to powers: } (-4)^2 = 16 \text{ and } 2^2 = 4.$$

$$= \frac{16 - 12}{-5 - (-3)} \qquad \text{Perform multiplication.}$$

$$= \frac{16 - 12}{-5 + 3} \qquad \text{Write subtraction as equivalent addition.}$$

$$= \frac{4}{-2} = -2 \qquad \text{Simplify both numerator and denominator, then divide.}$$

EXAMPLE 2   *Objective 2: Identify and evaluate algebraic expressions.*

Complete the table by evaluating the expression $\dfrac{10x^2 - 3x + 4}{2}$ for the given values of x.

x	1	2	3
$\dfrac{10x^2 - 3x + 4}{2}$			

Solution: To complete the table, evaluate $\dfrac{10x^2 - 3x + 4}{2}$ at each given replacement value.

When $x = 1$,

$$\frac{10x^2 - 3x + 4}{2} = \frac{10(1)^2 - 3(1) + 4}{2} = \frac{10(1) - 3(1) + 4}{2} = \frac{10 - 3 + 4}{2} = \frac{11}{2}.$$

When $x = 2$,

$$\frac{10x^2 - 3x + 4}{2} = \frac{10(2)^2 - 3(2) + 4}{2} = \frac{10(4) - 3(2) + 4}{2} = \frac{40 - 6 + 4}{2} = \frac{38}{2} = 19.$$

When $x = 3$,

$$\frac{10x^2 - 3x + 4}{2} = \frac{10(3)^2 - 3(3) + 4}{2} = \frac{10(9) - 3(3) + 4}{2} = \frac{90 - 9 + 4}{2} = \frac{85}{2}.$$

The completed table is:

x	1	2	3
$\dfrac{10x^2 - 3x + 4}{2}$	$\dfrac{11}{2}$	19	$\dfrac{85}{2}$

EXAMPLE 3   *Objective 3: Write word phrases as algebraic expressions.*

Write each as a mathematical expression.
a. The cost of y cans of soda if each can costs $0.32.
b. A piece of yarn is x centimeters long. If one portion is 15 centimeters long, represent the length of the other portion as an expression in x.

Solution: a.   The cost of the cans of soda is found by multiplying the cost of one can by the number of cans.

In words: | Cost of can | • | Number of cans |

Translate: 0.32   • y or $0.32y$

b. If a piece of yarn is x centimeters long and one portion is 15 centimeters long, the other portion is "the rest of x" or

In words: | x | | minus | | 15 |

Translate: x −   15   or $x - 15$

EXAMPLE 4   *Objective 4: Identify like terms and simplify algebraic expressions.*

Simplify each expression.
a. $4z + 3xy - 2 - 6z + 3 - xy$ b. $16y^2 - 3y + 5(y^2 + 2y)$ c. $(3b + 4c) - 7(1 + b - 2c)$
d. $3(7x - 2) + 5(y - 3x + 2)$

Solution: a.

$4z + 3xy - 2 - 6z + 3 - xy = 4z - 6z + 3xy - xy - 2 + 3$ Apply the commutative property.

$= (4 - 6)z + (3 - 1)xy + (-2 + 3)$ Apply the distributive property.

$= -2z + 2xy + 1$ Simplify.

b.

$16y^2 - 3y + 5(y^2 + 2y) = 16y^2 - 3y + 5y^2 + 10y$ Apply the distributive property.

$= 16y^2 + 5y^2 - 3y + 10y$ Apply the commutative property.

$= 21y^2 + 7y$ Combine like terms.

c.

$(3b + 4c) - 7(1 + b - 2c) = 3b + 4c - 7 - 7b + 14c$ Apply the distributive property.

$= -4b + 18c - 7$ Combine like terms.

d.

$3(7x - 2) + 5(y - 3x + 2) = 21x - 6 + 5y - 15x + 10$ Apply the distributive property.

$= 6x + 5y + 4$ Combine like terms.

EXERCISES

Exercises 1–4. Simplify each expression. See Example 1.

1. $3 \cdot 2^2 - 5$ **2.** $|7 - 10| + 4(-5)$ **3.** $\dfrac{15 + 3(-4)}{\sqrt{9}}$ **4.** $\dfrac{|3 - 4| + |-13|}{3^2 - 2}$

Exercises 5–8. Find the value of each expression when $x = 2$, $y = 10$, and $z = -1$. See Example 2.

5. $3x - y + 2z$ **6.** $8x^2 + 2(z - y)$ **7.** $-10(x + y + 5z)$ **8.** $\dfrac{2y}{x^2}$

9. The expression $\dfrac{24c}{a(p+1)}$ approximates the annual interest rate on a consumer loan. The finance charge is represented by c, the amount of the loan is represented by a, and the number of payments is represented by p.
  (a) Approximate the annual interest rate on a loan of $1000 that is to be paid off in 23 payments. The finance charge on the loan is $100.
  (b) Approximate the annual interest rate on a loan of $5000 that is to be paid off in 48 payments. The finance charge on the loan is $200.

10. The expression $2lw + 2hl + 2hw$ gives the surface area of a rectangular solid having length l, height h, and width w.
  (a) Find the surface area of a rectangular cardboard box that is 80 centimeters long, 30 centimeters high, and 40 centimeters wide.
  (b) Find the surface area of a rectangular cardboard box that is 50 inches long, 60 inches high, and 20 inches wide.

Exercises 11–12. Complete each table. See Example 2.

11.

x	−1	0	1	2
$2x^2 - 3$				

12.

y	−3	−1	2	4
$-5(4y + 10)$				

Exercises 13–20. Write each of the following as an algebraic expression. See Example 3.

13. The cost of x greeting cards if each card costs $1.65.
14. The cost of y gallons of gasoline if each gallon costs $1.179.
15. The cost of x apples and y oranges if oranges cost $0.39 each and apples cost $0.45 each.
16. The cost of c yards of fabric and d yards of ribbon if the fabric costs $3.49 per yard and the ribbon costs $0.49 per yard.
17. The sum of two numbers is 80. If one number is z, represent the other number as an expression in z.
18. The sum of two numbers is 243. If one number is q, represent the other number as an expression in q.
19. The length of a piece of fabric is 5.5 yards. If x yards must be cut from the fabric, represent the length of the piece that is left.
20. The length of a board is y. If 10 inches are cut off the board, represent the length of the piece that is left.

Exercises 21–28. Simplify each expression. See Example 4.

21. $2x - 5y + 4 - x + 10y$ **22.** $-a + 9(a + 3b)$ **23.** $2r(3 + r)$ **24.** $(t + 3) - (4t - 2)$

25. $-3(2u - 5) - 2(10u + 1)$ **26.** $4(n + 1) + 3^2(n - 1)$ **27.** $4x - 7y$ **28.** $-12b + 2(b - 6) - 5$

Exercises 29-34. Use the given line graph to answer each question. The graph gives the total number of passenger cars imported into the United States each year from 1989 through 1994. (data from U.S. Department of Commerce)

29. Estimate the number of cars imported into the United States in 1993.

30. Estimate the number of cars imported into the United States in 1994.

31. During which year was the greatest number of cars imported?

32. During which year was the least number of cars imported?

33. Estimate the difference in the number of imported cars for the years 1990 and 1992.

34. Estimate the difference in the number of imported cars for the years 1991 and 1993.

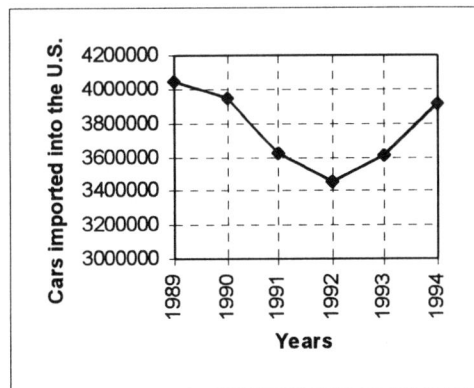

Chapter 1   **Hints and Warnings**

- To **evaluate** an algebraic expression containing variables, substitute the given numbers for the variables and simplify. *Take care when substituting negative numbers for variables.*
- To **add real numbers:**
  1) Two numbers with the same sign: add their absolute values and attach their common sign.
  2) Two numbers with different signs: subtract the smaller absolute value from the larger absolute value and attach the sign of the number with the larger absolute value. *Take care when adding two numbers with different signs. You may find checking yourself by using a number line helpful.*
- To **subtract real numbers,** rewrite as an equivalent addition problem and use the rules for adding real numbers. For example, rewrite $a - b$ as $a - b = a + (-b)$. *Remember that subtracting a negative number is like adding a positive number:* $a - (-b) = a + b$.
- To **multiply or divide real numbers:**
  1) A product or quotient of two numbers with the same sign is positive. *Remember that the product of several numbers with an even number of negative signs is also positive.*
  2) A product or quotient of two numbers with different signs is negative. *Remember that the product of several numbers with an odd number of negative signs is also negative.*

- To **simplify** an algebraic expression, use the order of operations given below. If grouping symbols such as parentheses are present, simplify expressions within those first, starting with the innermost set. If fraction bars are present, simplify the numerator and denominator separately.
    1) Raise to powers or take roots in order from left to right.
    2) Multiply or divide in order from left to right.
    3) Add or subtract in order from left to right.

CHAPTER 1 Practice Test

Determine whether each statement is true or false.

1. $-3.41 < -3.4$ **2.** $4^2 = (-4)^2$ **3.** $3 - 9 = -|3 - 9|$ **4.** $(6)(-1)(0) = \dfrac{6}{0}$

5. All whole numbers are natural numbers.   **6.** All irrational numbers are rational numbers.

Simplify.

7. $7 - 16 \div 4(2)$ **8.** $\left|1 - 3\right|^2 - (2 - 4^2)$ **9.** $(5 - 8)^3 - \left|-2 - 3\right|^2$

10. $[2 \cdot |8 - 9|^6 - (-7)] \div (-3)$ **11.** $\dfrac{4(11 - 9)^3 + (-8)}{(-2)(-3)(2)}$

Evaluate each expression when $x = -3$, $y = 0$, $z = 2$.

12. $x^3 + z^5$ **13.** $\dfrac{xy - z^2}{xz}$

14. The algebraic expression $24.95q$ represents the total revenue for q quick lube oil changes at a local chain.
(a) Complete the table below.

Oil changes	q	1	4	10	30
Revenue	$24.95q$				

(b) As the number of oil changes increases, does the total revenue increase or decrease?

Write each statement using mathematical symbols.

15. Three times the absolute value of the difference of 12 and z is 24.
16. The cube of the sum of x and seven, divided by five, is greater than or equal to zero.
17. Twice a increased by nine is not equal to the opposite of a.
18. Negative five is less than six times y divided by the absolute value of the sum of y and two.

Name each property illustrated.

19. $(-2 + p) + q = -2 + (p + q)$ **20.** $9 \cdot \dfrac{1}{9} = 1$

21. $3 \cdot x = x \cdot 3$ **22.** $3y + 6 = 3(y + 2)$

23. Write an expression for the total amount of money (in cents) in p pennies and q quarters.

CHAPTER 2

Equations, Inequalities and Problem Solving

2.1   Linear Equations in One Variable

EXAMPLE 1   *Objective 2: Solve linear equations.*

Solve for y: $53 - 7y = 4$.

Solution: First, use the addition property of equality and subtract 53 from both sides.

$$53 - 7y = 4$$
$$53 - 7y - 53 = 4 - 53$$
$$-7y = -49 \qquad \text{Simplify.}$$
$$\frac{-7y}{-7} = \frac{-49}{-7} \qquad \text{Divide both sides by } -7.$$
$$y = 7 \qquad \text{Simplify.}$$

To check, replace y in the original equation with 7.

$$53 - 7y = 4$$
$$53 - 7(7) = 4 \qquad \text{Let } y = 7.$$
$$53 - 49 = 4$$
$$4 = 4 \qquad \text{True.}$$

The solution set is $\{7\}$.

EXAMPLE 2   *Objective 3: Solve linear equations that can be simplified by combining like terms.*

Solve for z: $-4(z + 5) = 3(2z - 10)$.

Solution: First, use the distributive property on both sides.

$$-4(z + 5) = 3(2z - 10)$$
$$-4z - 20 = 6z - 30 \qquad \text{Use the distributive property.}$$
$$-4z - 20 - 6z = 6z - 30 - 6z \qquad \text{Subtract } 6z \text{ from both sides.}$$
$$-10z - 20 = -30 \qquad \text{Simplify.}$$
$$-10z - 20 + 20 = -30 + 20 \qquad \text{Add 20 to both sides.}$$
$$-10z = -10 \qquad \text{Simplify.}$$
$$\frac{-10z}{-10} = \frac{-10}{-10} \qquad \text{Divide both sides by } -10.$$
$$z = 1$$

Let $z = 1$ in the original equation to see that $\{1\}$ is the solution set.

EXAMPLE 3   *Objective 4: Recognize identities and equations with no solution.*

Solve for x: $5(2x - 1) = 10x + 3$.

Solution: First, use the distributive property and remove parentheses.

$$5(2x - 1) = 10x + 3$$
$$10x - 5 = 10x + 3 \qquad \text{Use the distributive property.}$$
$$10x - 5 - 10x = 10x + 3 - 10x \qquad \text{Subtract } 10x \text{ from both sides.}$$
$$-5 = 3 \qquad \text{Simplify.}$$

Because this statement is a contradiction, the solution set is $\varnothing$.

EXERCISES

Exercises 1–8. Solve each equation for the variable. See Example 1.

1. $12 - 2y = 6$ **2.** $3x - 5 = 40$ **3.** $8x + 8 = -16$ **4.** $7.5t - 42 = 37.2$

5. $55 = 4 + 17w$ **6.** $31.7 = -21.3 + 4r$ **7.** $-30 = -5q + 20$ **8.** $36 = 5x + 6$

Exercises 9–14. Solve each equation for the variable. See Example 2.

9. $10t - 3 = 5t + 12$ **10.** $12w - 16 = 13 + 4w$ **11.** $17 - 8y = -5y - 13$

12. $3(y + 2) = 14(y - 1)$ **13.** $-2(x - 6) = x + 15$ **14.** $-7(2x + 3) = 2(4 - 3x)$

Exercises 15–20. Decide whether each of the following equations has a solution, is an identity, or has no solution. See Example 3.

15. $3(5 - x) = 12 - x$ **16.** $7 - x = 2(0.5x + 2)$ **17.** $12(x - 3) = 4(3x - 9)$

18. $-2(5 - 2x) = 4(x - 3) + 2$ **19.** $9x - 4 = 5(x + 3) + 4x$ **20.** $6y - (2y + 1) = 4(y + 1) - 5$

Exercises 21–24. Solve each equation.

21. $\dfrac{1 + 2x}{3} = \dfrac{3 - 4x}{5}$ **22.** $\dfrac{5(x - 6)}{2} = \dfrac{x + 2}{6}$

23. $\dfrac{2y}{3} + \dfrac{y}{6} = 15$ **24.** $\dfrac{4y - 1}{10} - \dfrac{3y}{5} = 8$

Exercises 25–28. Write each statement as an algebraic expression. Then simplify.

25. The sum of three consecutive integers if the first integer is y.

26. The sum of three consecutive integers if the first integer is $3q$.

27. The total amount of money (in cents) in t dimes and $2t + 14$ pennies.

28. The total amount of money (in cents) in x nickels and $4(x+3)$ quarters.

2.2   An Introduction to Problem Solving

EXAMPLE 1   *Objective 1: Apply the steps for problem solving.*

Suppose that the price of a stock has increased 15% to $45 per share. What was the price of the stock before the price increased?

Solution: 1.   UNDERSTAND the problem. Read and reread the problem, and then propose a solution. For example, guess that the original price was $30. The amount of the increase is then 15% of $30, or $(0.15)(\$30) = \$4.50$. This means that the new price of the stock is the original price plus the increase, or $\$30 + \$4.50 = \$34.50$. Our guess is incorrect, but we now have a better idea of how to model the problem. We also know that the original price was greater than $30.

2.   ASSIGN a variable. Let x = the original price of the stock.

3.   ILLUSTRATE the problem. No illustration is needed here.

4.   TRANSLATE.

In words:

Original stock price	plus	15% of original price	is	new price

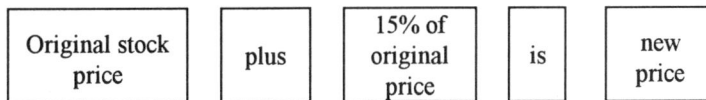

Translate: x $+$ $0.15x$ $=$ 45

5.   COMPLETE. Solve the equation.
$$x + 0.15x = 45$$
$$1.15x = 45$$
$$x = 45 / 1.15 \approx 39.13$$

6.   INTERPRET. *Check:* If the original price of the stock is $39.13, the new price is
$$\$39.13 + (0.15)(\$39.13) = \$39.13 + \$5.8695$$
$$= \$44.9995 \approx \$45, \text{ the given new price.}$$

State: The original price of the stock is $39.13.

EXAMPLE 2   *Objective 1: Apply the steps for problem solving.*

Find three consecutive integers such that the sum of the first number, three times the second number, and twice the third number is 49.

Solution: 1.   UNDERSTAND the problem. Read and reread the problem, and then propose a solution. For example, guess that the first number is 1, the second number is 2, and the third number is 3. This means that the sum that is described is $1 + 3(2) + 2(3) = 13$. Our guess is incorrect, but we now have a better idea of how to model this problem.

2.   ASSIGN a variable. Let x = the first number, then $x + 1$ = the second number, and $x + 2$ = the third number.

3.   ILLUSTRATE the problem. No illustration is needed here.

4. TRANSLATE.

In words:

| First number | plus | three times second number | plus | twice third number | is | 49 |

Translate: x $+$ $3(x+1)$ $+$ $2(x+2)$ $=$ 49

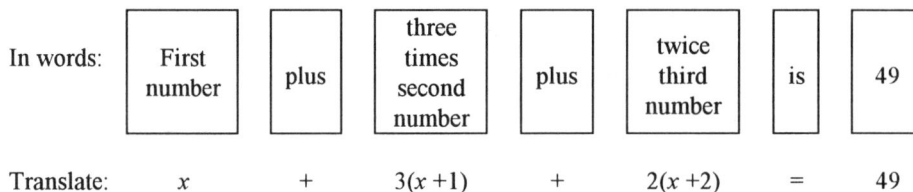

5. COMPLETE. Solve the equation.

$x + 3(x+1) + 2(x+2) = 49$

$x + 3x + 3 + 2x + 4 = 49$ Remove parentheses.

$6x + 7 = 49$ Combine like terms.

$6x = 42$ Subtract 7 from both sides

$x = 7$ Divide both sides by 6.

6. INTERPRET. If the first number is 7, then the second number is $7 + 1 = 8$, and the third number is $7 + 2 = 9$. *Check:*

$$7 + 3(8) + 2(9) = 7 + 24 + 18$$
$$= 49, \text{ the given sum.}$$

State: The three consecutive integers are 7, 8, and 9.

EXERCISES

Exercises 1–2. Solve. See Example 1.

1. An advertisement states that the regular price of a car will be discounted by 10% for a limited time. The car's sale price is $14,395.50. What is the regular price of the car?

2. A stamp in a stamp collector's collection has increased in value by 30% over the past ten years. The current value of the stamp is $29.90. What was the value of the stamp ten years ago?

Exercises 3–8. Solve.

3. Ten years ago, a house was worth $80,000. Today its value has increased by 15%. What is its value today?

4. The original price of a new car was $7000 twelve years ago. Since then, the value of the car has decreased 65%. What is the car worth now?

5. Public elementary and secondary schools in Texas receive 43.1% of their funding from the state. These schools in Texas received a total of $19,930,447 in funding for the 1994-1995 school year. How much money did the schools receive from the state of Texas? (Source: National Education Association)

6. Public elementary and secondary schools in Louisiana receive 12.4% of their funding from the federal government. These schools in Louisiana received a total of $3,768,960 in funding for the 1994-1995 school year. How much money did the schools receive from the federal government? (Source: National Education Association)

7. The average size of a farm in the United States has increased by 170.7% from 1940 to 1994. In 1994 the average size of a farm in the United States was 471 acres. What was the average size of a farm in the United States in 1940? Round to the nearest whole number. (data from U.S. Department of Agriculture)

8. In 1994, labor union members made up 15.5% of the United States labor force. This represents a total labor union membership of 16,748,000. How many workers were there in the entire United States labor force in 1994? Round to the nearest whole number. (data from Bureau of Labor Statistics)

Exercises 9–10. Solve. See Example 2.

9. Find three consecutive integers such that their sum is 141.

10. Find three consecutive integers such that the sum of the second integer, the third integer, and twice the first integer is 63.

Exercises 11–15. Solve.

11. The years in which the television show "Bonanza" was rated number one are three consecutive integers. If the sum of the first integer, twice the second integer, and the third integer is 7860, find each year. (data from Nielsen Media Research)

12. Five times the sum of a number and 6 is the same as 3 times the difference of 12 and the number. Find the number.

13. The sum of the angles in a triangle is 180°. Find the angles of a triangle whose two base angles are equal, and whose third angle is 15° more than a base angle.

14. Find an angle such that its complement is equal to one-third of the sum of its supplement and 18°.

15. A university drama department staged a production. Ticket sales were categorized by type of play attendee. The results are summarized in the pie chart.
    (a) What percent of ticket sales were made to members of the community?
    (b) What category accounted for the largest portion of ticket sales?
    (c) What category accounted for the smallest portion of ticket sales?
    (d) If 300 tickets were sold, how many were sold to engineering students?
    (e) If 300 tickets were sold, how many were sold to members of the community?

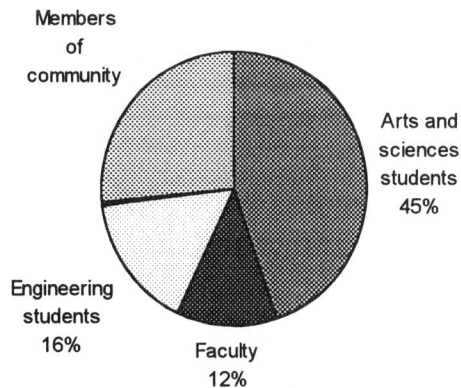

Members of community

Arts and sciences students 45%

Engineering students 16%

Faculty 12%

2.3   Formulas and Problem Solving

EXAMPLE 1   *Objective 1: Solve a formula for a specified variable.*

Solve $5(2w + 3z) = -20$ for z.

Solution: This is a linear equation in two variables. We begin with step 2 in how "To Solve Equations for a Specified Variable" and use the distributive property to remove the parentheses.

$$5(2w + 3z) = -20$$

$$10w + 15z = -20 \qquad \text{Apply the distributive property.}$$

$$10w + 15z - 10w = -20 - 10w \qquad \text{Subtract } 10w \text{ from both sides.}$$

$$15z = -20 - 10w \qquad \text{Simplify.}$$

$$\frac{15z}{15} = \frac{-20 - 10w}{15} = \frac{-20}{15} + \frac{-10w}{15} \qquad \text{Divide both sides by 15.}$$

$$z = -\frac{4}{3} - \frac{2w}{3} \qquad \text{Simplify.}$$

EXAMPLE 2   *Objective 2: Use formulas to solve problems.*

Suppose you are baking a pie. To keep the edges of the pie crust from burning, you would like to wrap a one-inch strip of aluminum foil around the edge of the pan. If you are using a round pie pan that is 9 inches in diameter, how long should be the strip of aluminum foil?

Solution: 1.   UNDERSTAND the problem. Read and reread the problem. The appropriate formula needed to solve this problem is the circumference of a circle formula $C = 2\pi r$.

2.   ASSIGN a variable. Let C = the circumference of the pie pan, and r = the radius of the pie pan.

3.   ILLUSTRATE the problem.

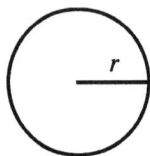

4.   TRANSLATE. Use the circumference formula and substitute known values. We see that the information we are given in the problem doesn't quite fit the formula because we were given the diameter, and not the radius, of the pan. However, we know that the radius is half of the diameter, so $r = 4.5$ inches.

   Formula: $C = 2\pi r$

   Substitute: $C = 2\pi(4.5)$

5.   COMPLETE. Here, we simplify the right side of the equation.

$$C = 2\pi(4.5)$$

$$= 9\pi \qquad \text{Multiply.}$$

$$\approx 28.27433388 \qquad \text{Approximate } \pi.$$

6. INTERPRET. To check here, repeat your calculations to make sure that no error was made. *State:* The strip of aluminum foil should be at least 28.3 inches long.

EXERCISES

Exercises 1–6. Solve each equation for the specified variable. See Example 1.

1. $D = rt$; for r

2. $3y - 10x = 27$; for y

3. $A = 3M - 2N$; for M

4. $E = I(r + R)$; for R

5. $A = 5H(b + B)$; for H

6. $N = 3st^4 - 5sv$; for s

Exercises 7–8. Solve. See Example 2

7. What is the circumference of a pizza with a diameter of 12 inches?

8. A circular rug with a diameter of 5 feet must be bound with bias tape around its outer edge to keep it from fraying. What length of bias tape is required?

Exercises 9–20. Solve.

9. You have deposited $2000 in a savings account that pays 4% compounded quarterly. If no other deposits are made, how much money will be in the account after 2 years?

10. You have invested $1000 in an account paying 7% interest compounded monthly. If no other deposits are made, how much money will be in the account after 1 year?

11. You have deposited $5000 in an account paying 3.5% interest compounded every two weeks. If no other deposits are made, how much money will be in the account after 1 year?

12. You have invested $7000 in a certificate of deposit paying 5.2% interest compounded every two weeks. How much money will be in the account after 3 years?

13. A patchwork quilt measuring 90 inches by 66 inches is to be made out of 6-inch square fabric patches.
(a) How many patches are needed to make the quilt?
(b) How many rows of patches are needed?
(c) How many patches should be in each row?

14. A package of floor tiles contains 144 two-inch-square tiles. You will use the tiles to floor a 6-feet by 8-feet rectangular utility room.
(a) How many tiles are needed?
(b) How many packages of tiles should you buy?

15. Suppose you are traveling at a constant speed of 62 miles per hour. You are 170 miles away from Cleveland, Ohio. Can you reach Cleveland in less than 3 hours?

16. Chicago, Illinois, is 202 miles from Springfield Illinois. Suppose you made this trip in 3.5 hours. What was your average speed?

17. The longest bridge span in the United States is the Verrazano-Narrows bridge in New York City. It is 4260 feet long (there are 5280 feet in one mile). If you are traveling at a speed of 45 miles per hour, how long will it take to drive over the bridge? (data from Survey of State Highway Engineers)

18. The Seikan tunnel in Japan is the world's longest railroad tunnel. It is 33.5 miles long. If a bullet train travels at a speed of 95 miles per hour, how long will the train's engine be in the tunnel? (data from *Railway Directory & Year Book*)

19. Suppose you have placed three golf balls in a paper bag. One ball is painted red, one green, and one yellow. You draw one ball out of the bag at random.
  (a) What is the probability that the ball drawn from the bag is green?
  (b) What is the probability that the ball drawn from the bag is yellow?
  (c) What is the probability that the ball drawn from the bag is **not** yellow?

20. Suppose you place slips of paper numbered from 1 through 10 in a bag and draw one slip at random.
  (a) What is the probability that the number drawn is 6?
  (b) What is the probability that the number drawn is odd?
  (c) What is the probability that the number drawn is 1, 2, or 3?

2.4   Linear Inequalities and Problem Solving

EXAMPLE 1   *Objective 2: Use interval notation.*

Write each set in interval notation.
  a. $\{x|x \le -5\}$ **b.** $\{x|x > -32\}$ **c.** $\{x|x < 7\}$ **d.** $\{x|-2 \le x < 1\}$

Solution: **a.** $(-\infty, -5]$ **b.** $(-32, \infty)$ **c.** $(-\infty, 7)$ **d.** $[-2, 1)$

EXAMPLE 2   *Objective 3: Solve linear inequalities using the addition property of inequality.*

Solve for y: $7y - 17 \le 6y - 8$. Write the solution in interval notation.

Solution: Start by grouping the variable terms on one side.

$$7y - 17 \le 6y - 8$$

$7y - 17 - 6y \le 6y - 8 - 6y$	Subtract $6y$ from both sides.
$y - 17 \le -8$	Simplify.
$y - 17 + 17 \le -8 + 17$	Add 17 to both sides.
$y \le 9$	Simplify.

The solution set is $\{y|y \le 9\}$, which in interval notation is $(-\infty, 9]$.

EXAMPLE 3   *Objective 4: Solve linear inequalities using the multiplication property of inequality.*

Solve for x: $3(3 - x) - 1 \le x - 4$. Write the solution in interval notation.

Solution: Start by applying the distributive property.

$$3(3 - x) - 1 \le x - 4$$

$9 - 3x - 1 \le x - 4$	Apply the distributive property.
$8 - 3x \le x - 4$	Combine like terms.

$$8 - 4x \leq -4 \qquad \text{Subtract } x \text{ from both sides.}$$
$$-4x \leq -12 \qquad \text{Subtract 8 from both sides.}$$
$$\frac{-4x}{-4} \geq \frac{-12}{-4} \qquad \begin{array}{l}\text{Divide both sides by } -4 \text{ and reverse} \\ \text{the inequality symbol.}\end{array}$$
$$x \geq 3 \qquad \text{Simplify.}$$

The solution set written in interval notation is $[3, \infty)$.

EXERCISES

Exercises 1–8. Write each set in interval notation. See Example 1.

1. $\{x | x > 28\}$ **2.** $\{y | y < -5\}$ **3.** $\{y | y \leq 16\}$ **4.** $\{x | x \geq -1\}$

5. $\{x | -3 \leq x < 11\}$ **6.** $\{t | 9 \leq t \leq 17\}$ **7.** $\{x | -4 < x \leq -2\}$ **8.** $\{y | -18 < y < 8\}$

Exercises 9–17. Solve each inequality for the variable. Write the solution in interval notation. See Examples 2–3.

9. $6 + 3x \leq 3$ **10.** $2x - 7 < x + 8$ **11.** $6 - 2x \geq 2$

12. $5 - x \leq 3x - 7$ **13.** $5(x + 2) > 2(2x - 1)$ **14.** $-2(x - 2) \geq 4(3 - x)$

15. $\dfrac{5x + 15}{4} < 0$ **16.** $\dfrac{-2x + 9}{3} \leq 7$ **17.** $\dfrac{1}{2}(2x - 5) > 2 - x$

18. Suppose you have scores of 83, 92, 85, and 85 on your biology tests. The final exam in biology will count as two tests. Use an inequality to find the minimum score you can make on the final exam to receive a course average of at least 85 (the cut-off for a grade of B+).

19. An algebra student has three test scores of 65, 82, and 93. In order to advance to the next math course in the sequence, her final grade must be at least 75. If the final exam counts as three tests, use an inequality to find the minimum score she can make on the final exam to advance to the next math course.

20. To send a package FedEx Standard Overnight®, FedEx charges $15 for the first pound, $1 for each additional pound up to a total of 5 pounds, and a $2.50 courier pick-up fee.
   (a) Use an inequality to find the maximum weight that can be shipped overnight for $19.
   (b) If you drop off the package instead of using a courier, you can avoid the $2.50 fee. Use an inequality to find the maximum weight that can be shipped overnight for $19 if you drop off the package.

2.5   Compound Inequalities

EXAMPLE 1   *Objective 1: Solve compound inequalities.*

Solve for y: $3 \leq -5 + 2y < 11$

Solution: To isolate y, first add 5 to all three sides.

$$3 \leq -5 + 2y < 11$$

$$3 + 5 \leq -5 + 2y + 5 < 11 + 5 \qquad \text{Add 5 to all three sides.}$$

$$8 \leq 2y < 16 \qquad\qquad\qquad \text{Simplify.}$$

$$\frac{8}{2} \leq \frac{2y}{2} < \frac{16}{2} \qquad\qquad \text{Divide all three sides by 2.}$$

$$4 \leq y < 8 \qquad\qquad\qquad \text{Simplify.}$$

The solution set in interval notation is $[4, 8)$.

EXAMPLE 2   *Objective 1: Solve compound inequalities.*

Solve for x: $8x + 3 \geq 35$ or $x + 12 < 6$.

Solution: Solve each inequality separately.

$$8x + 3 \geq 35 \qquad\qquad x + 12 < 6$$

$$8x \geq 32$$

$$x \geq 4 \qquad \text{or} \qquad x < -6$$

The solution is the union of these intervals. The solution set in interval notation is $(-\infty, -6) \cup [4, \infty)$.

EXAMPLE 3   *Objective 1: Solve compound inequalities.*

Solve for x: $\frac{1}{4}x - 1 < 10$ or $4 - \frac{1}{2}x < 9$.

Solution: Solve each inequality separately.

$$\frac{1}{4}x - 1 < 10 \qquad\qquad\qquad 4 - \frac{1}{2}x < 9$$

$$\frac{1}{4}x - 1 + 1 < 10 + 1 \qquad\qquad 4 - \frac{1}{2}x - 4 < 9 - 4$$

$$\frac{1}{4}x < 11 \qquad\qquad\qquad -\frac{1}{2}x < 5$$

$$(4)\left(\frac{1}{4}x\right) < (4)11 \qquad\qquad (-2)\left(-\frac{1}{2}x\right) > (-2)5$$

$$x < 44 \qquad \text{or} \qquad x > -10$$

The solution is the union of these intervals. The solution set in interval notation is $(-\infty, \infty)$.

EXERCISES

Exercises 1–8. If $A = \{x | x$ is a nonnegative integer$\}$, $B = \{x | x$ is a negative integer$\}$, $C = \{-3, -2, -1, 0\}$, and $D = \{0, 1, 2, 3\}$, list the elements of each set.

1. $A \cup B$
2. $A \cap B$
3. $C \cup D$
4. $C \cap D$
5. $A \cap C$
6. $A \cap D$
7. $B \cap D$
8. $B \cap C$

Exercises 9–14. Solve each compound inequality for the variable. Write the solution in interval notation. See Example 1.

9. $6 + 3x \ge 3$ and $x - 6 < 0$

10. $2x - 7 < x$ and $3x > -2$

11. $9 \le x - 25 \le 14$

12. $4 > 2(x - 1) > -5$

13. $-12 \le 4(2 - x) < 17$

14. $14 < \dfrac{1}{2}(1 - 6x) \le 33$

Exercises 15–20. Solve each compound inequality for the variable. Write the solution in interval notation. See Examples 2 and 3.

15. $x \le 0$ or $7x \ge 49$

16. $y > -2$ or $y + 3 < -8$

17. $4x < 2$ or $x - 5 \ge 12$

18. $x + 6 > 21$ or $\dfrac{1}{3}x \le 4$

19. $-2.7y < 4$ or $5.6y < 30$

20. $4.2x - 6 \le 20.3$ or $12.9 > 3 - \dfrac{1}{2}x$

Exercises 21–22. Use the formula $F = \dfrac{9}{5}C + 32$ to convert Celsius temperatures to Fahrenheit temperatures.

21. In 1994, the temperatures in Jacksonville, Florida, ranged from 26°F to 96°F. Use a compound inequality to convert this temperature range to the Celsius scale. (data from National Climatic Data Center)

22. In 1991, the temperatures in Rapid City, South Dakota, ranged from – 25°F to 103°F. Use a compound inequality to convert this range to the Celsius scale. (data from National Climatic Data Center)

23. A student has scores of 94, 69, 77 on his French tests. Use a compound inequality to find the range of scores he can make on the final exam to receive a B in the course. The final exam counts as two tests, and a B is received if the final course average is at least 82 and less than 86.

24. A student has test scores of 75, 81, and 80. Use a compound inequality to decide if it is possible for her to receive an A– in the course if the final exam counts as three tests. An A– is received if the final course average is at least 90 and less than 94.

25. The formula $C = 14 + 8.5x$ gives the cost of ordering x reams of paper. Your company would like to spend an amount in the range $200 \le C \le \$350$ on paper. Use a compound inequality to find the corresponding interval of number of reams of paper that could be purchased.

26. The formula $C = 500 + 4.25x$ gives the cost of making x T-shirts. To make a profit, your company must spend an amount in the range $1000 \le C \le \$4000$. Use a compound inequality to find the corresponding interval of number of T-shirts that could be made.

2.6   Absolute Value Equations

EXAMPLE 1   *Objective 1: Solve absolute value equations.*

Solve for y: $|9 - 4.5y| = 18$

Solution: Here the expression inside the absolute value bars is $9 - 4.5y$. If we think of the expression $9 - 4.5y$ as x in the absolute value property, we have $|x| = 18$, which is equivalent to $x = 18$ or $x = -18$. By substituting $9 - 4.5y$ for x, we solve the following two equations for y.

$$9 - 4.5y = 18 \qquad\qquad 9 - 4.5y = -18$$
$$9 - 4.5y - 9 = 18 - 9 \quad\text{or}\quad 9 - 4.5y - 9 = -18 - 9$$
$$-4.5y = 9 \quad\text{or}\quad -4.5y = -27$$
$$y = -2 \quad\text{or}\quad y = 6$$

To check, let $y = -2$ and then $y = 6$ in the original equation.

$$\text{Let } y = -2 \qquad\qquad \text{Let } y = 6$$
$$|9 - 4.5(-2)| = 18 \qquad |9 - 4.5(6)| = 18$$
$$|9 + 9| = 18 \qquad |9 - 27| = 18$$
$$|18| = 18 \qquad |-18| = 18$$
$$18 = 18 \quad\text{True} \qquad 18 = 18 \quad\text{True}$$

Both solutions check and the solution set is $\{-2, 6\}$.

EXAMPLE 2   *Objective 1: Solve absolute value equations.*

Solve for x: $14 - |5x| = -6$

Solution: We want the absolute value expression alone on one side of the equation, so begin by subtracting 14 from both sides. Then apply the absolute value property.

$$14 - |5x| = -6$$
$$14 - |5x| - 14 = -6 - 14 \qquad \text{Subtract 14 from both sides.}$$
$$-|5x| = -20 \qquad \text{Simplify.}$$
$$|5x| = 20 \qquad \text{Multiply both sides by } -1.$$
$$5x = 20 \quad\text{or}\quad 5x = -20$$
$$x = 4 \quad\text{or}\quad x = -4$$

The solution set is $\{-4, 4\}$.

EXAMPLE 3   *Objective 1: Solve absolute value equations.*

Solve for x: $\left|\dfrac{1}{2}x - 3\right| = |3x + 2|$.

Solution: This equation is true if the expressions inside the absolute value bars are equal or are opposites of each other.

$$\frac{1}{2}x - 3 = 3x + 2 \qquad\text{or}\qquad \frac{1}{2}x - 3 = -(3x + 2)$$

Next, solve each equation.

$$\frac{1}{2}x - 3 = 3x + 2 \quad \text{or} \quad \frac{1}{2}x - 3 = -3x - 2$$

$$-\frac{5}{2}x - 3 = 2 \quad \text{or} \quad \frac{7}{2}x - 3 = -2$$

$$-\frac{5}{2}x = 5 \quad \text{or} \quad \frac{7}{2}x = 1$$

$$x = 5\left(-\frac{2}{5}\right) = -2 \quad \text{or} \quad x = \frac{2}{7}$$

The solution set is $\left\{-2, \frac{2}{7}\right\}$.

EXERCISES

Exercises 1–8. Solve each absolute value equation. See Example 1.

1. $|w| = 27$ **2.** $|y| = 6.5$ **3.** $|5x| = 45$ **4.** $|12y| = 42$

5. $|10x - 6| = 54$ **6.** $|7 - 2x| = 23$ **7.** $|4 - 2x| = 12$ **8.** $|3x + 8| = 16$

Exercises 9–12. Solve each absolute value equation. See Example 2.

9. $|x| - 7 = -3$ **10.** $|x| + 2 = 71$ **11.** $|2x| - 4 = 35$ **12.** $|6x| + 12 = 84$

Exercises 13–16. Solve each absolute value equation. See Example 3.

13. $|x + 2| = |2x - 5|$ **14.** $|3x - 1| = |x + 9|$ **15.** $|2x + 11| = |3x - 4|$ **16.** $|5x - 7| = |4x - 13|$

Exercises 17–24. Solve each absolute value equation.

17. $|2x - 3| - 7 = 6$ **18.** $|7 - 5x| + 4 = 46$ **19.** $|-19| = |2x + 1|$ **20.** $|6x - 5| = |-38|$

21. $|x + 7| = |2x - 14|$ **22.** $|3x + 5| = |4x + 12|$ **23.** $\left|\frac{x + 3}{4}\right| = 16$ **24.** $\left|\frac{2y - 1}{3}\right| = |-7|$

25. The circle graph at the right shows the responses to a survey of owners of new domestic and foreign cars when asked how long they expected to own their new cars. Use this graph to answer the following questions. (data from J.D. Powers and Associates)

   (a) What percent of the survey respondents expected to own their cars for three years?

   (b) What percent of survey respondents expected to own their cars for either four or five years?

   (c) If a circle consists of 360°, find the number of degrees in the sector representing those who expect to own their new cars for four years.

   (d) If 500 people responded to the survey, how many of them expect to own their cars for six to eight years?

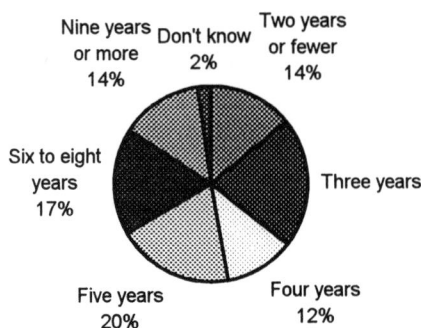

Nine years or more 14% — Don't know 2% — Two years or fewer 14% — Six to eight years 17% — Three years — Five years 20% — Four years 12%

2.7   Absolute Value Inequalities

EXAMPLE 1   *Objective 1: Solve absolute value inequalities of the form $|x| < a$.*

Solve for b: $|8 - 3b| - 7 \le 5$

Solution: First, isolate the absolute value expression by adding 7 to both sides.

$$|8 - 3b| - 7 \le 5$$
$$|8 - 3b| \le 5 + 7 \quad \text{Add 7 to both sides.}$$
$$|8 - 3b| \le 12 \quad \text{Simplify.}$$

Because 12 is positive, we apply the absolute value property for $|x| < a$.

$$-12 \le 8 - 3b \le 12$$
$$-12 - 8 \le 8 - 3b - 8 \le 12 - 8 \quad \text{Subtract 8 from all three sides.}$$
$$-20 \le -3b \le 4 \quad \text{Simplify.}$$
$$\frac{-20}{-3} \ge \frac{-3b}{-3} \ge \frac{4}{-3} \quad \text{Divide all three sides by } -3 \text{ and reverse the inequality symbols.}$$
$$\frac{20}{3} \ge b \ge -\frac{4}{3} \quad \text{Simplify.}$$

The solution set is $\left[-\frac{4}{3}, \frac{20}{3} \right]$.

EXAMPLE 2   *Objective 1: Solve absolute value inequalities of the form $|x| < a$.*

Solve for x: $14 + |5x| < 9$

Solution: First, isolate the absolute value expression on one side of the inequality by subtracting 14 from both sides.

$$14 + |5x| < 9$$
$$|5x| < 9 - 14 \qquad \text{Subtract 14 from both sides.}$$
$$|5x| < -5 \qquad \text{Simplify.}$$

The absolute value of a number is always nonnegative and can never be less than –5. Thus, this absolute value inequality has no solution. The solution set is $\varnothing$.

EXAMPLE 3   *Objective 2: Solve absolute value inequalities of the form* $|x| > a$.

Solve for y: $|4y - 3| + 10 \geq 73$.

Solution: First, isolate the absolute value expression by subtracting 10 from both sides.
$$|4y - 3| + 10 \geq 73$$
$$|4y - 3| + 10 - 10 \geq 73 - 10$$
$$|4y - 3| \geq 63$$

Next, write the absolute value inequality as an equivalent compound inequality and solve.

$4y - 3 \leq -63$	or	$4y - 3 \geq 63$
$4y \leq -60$	or	$4y \geq 66$
$y \leq -15$	or	$y \geq 16.5$

The solution set is $(-\infty, -15] \cup [16.5, \infty)$.

EXERCISES

Exercises 1–4. Solve each inequality. See Example 1.

1. $|w| < 27$ **2.** $|y| \leq 6.5$ **3.** $|4 - 2x| \leq 12$ **4.** $|3x + 8| < 16$

Exercises 5–8. Solve each inequality. See Example 3.

5. $|5x| \geq 45$ **6.** $|12y| > 42$ **7.** $|10x - 6| > 54$ **8.** $|7 - 2x| \geq 23$

Exercises 9–18. Solve each absolute value equation.

9. $|x + 2| > 23$ **10.** $|3x - 1| \geq |-19|$ **11.** $|2x + 11| \leq -4$ **12.** $|5x - 7| < -20$

13. $|2x - 3| - 7 < 6$ **14.** $|7 - 5x| + 4 \leq |-56|$ **15.** $|2(x + 3)| > 12$ **16.** $|4(x - 2)| + 3 \leq 7$

17. $\left|\dfrac{w-5}{3}\right| \geq 8$

18. $\left|\dfrac{-3(x+3)}{7}\right| < 5$

Exercises 19–20. The expression $\left|X_T - X\right|$ is defined to be the absolute error in X where X_T is the true value of a quantity and X is the measured value or the value as stored in a computer.

19. If the exact value of a quantity is $\dfrac{63}{37}$ and the absolute error must be less than or equal to 0.05, find the possible measured values.

20. If the exact value of a quantity is 0.079 and the absolute error must less than 0.001, find the possible measured values.

21. Suppose you have placed five slips of paper, each with one of the letters A, P, P, L, and E on it, into a hat. Find the probability of each situation when a single piece of paper is drawn from the hat at random.
    (a) P(drawing an A)   (b) P(drawing a P)

    (c) P(drawing a vowel)   (d) P(drawing a consonant)

Chapter 2   Hints and Warnings

- To **solve a linear equation in one variable:**
  1) Clear the equation of fractions. *Remember that this can often be done by multiplying by the least common multiple.*
  2) Remove grouping symbols such as parentheses. *Don't forget to apply the distributive property where appropriate.*
  3) Simplify by combining like terms.
  4) Write variable terms on one side and numbers on the other side using the addition property of equality. *Remember that the addition property of equality guarantees that the same number may be added to or subtracted from both sides of an equation and the result is an equivalent equation.*
- To **solve a problem,** use the following steps:
  1) Understand the problem.
  2) Assign a variable.
  3) Illustrate the problem.
  4) Translate the problem.
  5) Complete the problem by solving.
  6) Interpret the result.
  Don't forget that there are often many different ways to solve the same problem. Don't be afraid to be creative or use several different approaches. If you aren't sure where to start, you can always try guessing a solution and checking it. Even if the guess isn't correct, you will at least have a better understanding of how to approach the problem.

- To **solve a formula for a specified variable,** use the steps for solving an equation. Treat the specified variable as the only variable in the equation. *Be sure to combine any like terms after solving for the specified variable.*
- To **solve a linear inequality in one variable:**
  1) Clear the inequality of fractions.
  2) Remove grouping symbols such as parentheses.
  3) Simplify by combining like terms.
  4) Write variable terms on one side of the inequality and numbers on the other side using the addition property of inequality.
  5) Isolate the variable using the multiplication property of inequality. *Remember that whenever you multiply or divide both sides of an inequality by a negative number you must reverse the direction of the inequality sign.*
- To **solve an absolute value equation,** rewrite the equation $|x| = a$ as two separate equations $x = a$ or $x = -a$.
- To **solve an absolute value inequality:**
  1) Rewrite $|x| < a$ as $-a < x < a$ and solve.
  2) Rewrite $|x| > a$ as $x < -a$ or $x > a$ and solve.

CHAPTER 2   Practice Test

Solve each equation.

1. $9x + 22 = 4x - 18$

2. $4(x - 2) = 17 - 3(4 - x)$

3. $-2(y - 6) + 3y = 4\left(5 - \dfrac{3y}{4}\right)$

4. $5(n + 3) + 2n = 7(n + 2) + 1$

5. $\dfrac{h}{3} + \dfrac{h}{6} = 13$

6. $\dfrac{3z}{5} - 7 = \dfrac{7z}{10} + 8$

7. $|5 - 2x| = 3$

8. $|16r - 5| = -1$

Solve each equation for the specified variable.

9. $10x - 14y = 28;\ y$

10. $6(3r - 5q) - 5(2r + 3q) = 0;\ r$

11. $S = gt^2 + gvt;\ v$

12. $V = lwh;\ w$

Solve each inequality.

13. $4(x-5) - 3x < 2(x-8)$ **14.** $-6 + \dfrac{x}{3} \geq 2$ **15.** $4 \leq 3x + 4 < 13$ **16.** $|2x - 3| > 7$

17. $x \leq -1$ and $x \leq -6$ **18.** $x \leq -1$ or $x \leq -6$

19. $x < 2x - 3$ and $2(x-2) \leq 2(10 - x) + 20$ **20.** $7x - 3 > 18$ or $2x - 13 < 3$

21. Find 22% of 95.

22. As of May 1996, 21 states had raised speed limits from 55 mph on interstate highways outside urban areas following the passing of a federal law eliminating the mandated 55 mph speed limit in December 1995. What percent of all states had raised their speed limits? (data from Advocates for Highway Safety)

23. A circular window has a circumference of 125.6 inches. You plan to cover the window with heavy plastic during the winter to seal out drafts. If you plan to use a square piece of plastic whose width is equal to the diameter of the window, how large must be the square of plastic?

24. A company that makes board games figures the cost C to make x games weekly is given by $C = 2449 + 3.3x$, and the weekly revenue R is given by $R = 9.5x$. Use an inequality to find the number of games that must be sold to make a profit. (Revenue must exceed cost in order to make a profit.)

25. Find the amount of money in an account after 5 years if a principal of $5000 was invested at 4.2% interest compounded quarterly and no withdrawals were made. (Round to the nearest cent.)

CHAPTER 3

Graphs and Functions

3.1   Graphing Equations

EXAMPLE 1   *Objective 2: Determine whether an ordered pair of numbers is a solution to an equation in two variables.*

Determine whether (–4, 19) or (2, 0) are solutions of the equation $7x + 2y = 10$.

Solution: To check each ordered pair, replace x with the x-coordinate and y with the y-coordinate and see whether a true statement results.

Let $x = -4$ and $y = 19$
$$7x + 2y = 10$$
$$7(-4) + 2(19) = 10$$
$$-28 + 38 = 10$$
$$10 = 10 \quad \text{True.}$$

Let $x = 2$ and $y = 0$
$$7x + 2y = 10$$
$$7(2) + 2(0) = 10$$
$$14 + 0 = 10$$
$$14 = 10 \quad \text{False.}$$

Thus, (–4, 19) is a solution and (2, 0) is not a solution.

EXAMPLE 2   *Objective 3: Graph linear equations.*

Graph the equation $-3x + 2y = 11$.

Solution: Find three ordered pair solutions, and plot the ordered pairs. The line through the plotted points is the graph. Let's let x be 0, x be 1, and y be 0 to find our three ordered pairs.

Let $x = 0$
$$-3x + 2y = 11$$
$$-3(0) + 2y = 11$$
$$2y = 11$$
$$y = \frac{11}{2}$$

Let $x = 1$
$$-3x + 2y = 11$$
$$-3(1) + 2y = 11$$
$$2y = 14$$
$$y = 7$$

Let $y = 0$
$$-3x + 2y = 11$$
$$-3x + 2(0) = 11$$
$$-3x = 11$$
$$x = -\frac{11}{3}$$

The three ordered pairs $\left(0, \frac{11}{2} \right)$, (1, 7), $\left(-\frac{11}{3}, 0 \right)$ are listed vertically in the table below, and the graph of $-3x + 2y = 11$ is shown.

x	y
0	$\dfrac{11}{2}$
1	7
$-\dfrac{11}{3}$	0

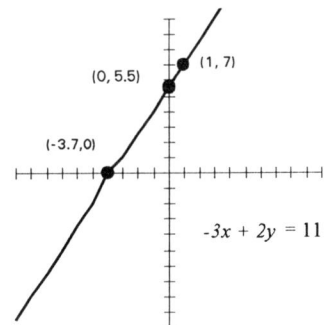

EXERCISES

Exercises 1–4. Determine whether each ordered pair is a solution of the given equation. See Example 1.

1. $x + y = 8$; (2, 5), (–1, 9)

2. $-8x + 3y = -6$; (0, –2), (3, 8)

3. $-5x + y = 11$; (3, 26), (–2, 0)

4. $x - 2y = 8$; (–1, 5), (8, 0)

Exercises 5–12. Graph each equation. See Example 2.

5. $x + y = 7$

6. $x - y = 3$

7. $3x - y = 8$

8. $\dfrac{3}{2}x + y = 2$

9. $2x + 2y = 5$

10. $4x - 2y = -9$

11. $y = \dfrac{3}{4}x - 3$

12. $y = -\dfrac{1}{2}x + 4$

C *Exercises 13–16. Use a grapher's standard window to graph the following equations.*

13. $y = 0.25x - 3.75$

14. $y = -0.33x + 5$

15. $y = 5x + 9$

16. $y = -4x - 7$

3.2   Introduction to Functions

EXAMPLE 1   *Objective 2: Identify functions.*

Which of the following relations are also functions?
a. $\{(2, 3), (5, 0), (2, 1), (10, 17)\}$ b. $\{(10, 2), (15, 3), (40, 8)\}$

c. $x = y^3$ d. $x = \dfrac{1}{2}y^8$

Solution: a.   Because the ordered pairs (2, 3) and (2, 1) belong to this relation, thus giving two different *y*-values for the same *x*-value, this relation is not a function.
b.   Each *x*-value is assigned to only one *y*-value. Thus, this relation is a function.
c.   The relation $x = y^3$ is a function if each *x*-value corresponds to just one *y*-value. For each *y*-value only one *x*-value is possible. Thus, $x = y^3$ is a function.
d.   In $x = \dfrac{1}{2}y^8$, if $y = -1$, then $x = \dfrac{1}{2}$. Also, if $y = 1$, then $x = \dfrac{1}{2}$. Thus the *x*-value $\dfrac{1}{2}$ corresponds to two different *y*-values, −1 and 1. Thus, $x = \dfrac{1}{2}y^8$ is not a function.

EXAMPLE 2   *Objective 3: Use the vertical line test for functions.*

Which of the following graphs are graphs of functions?
a.   b.   c.   d.

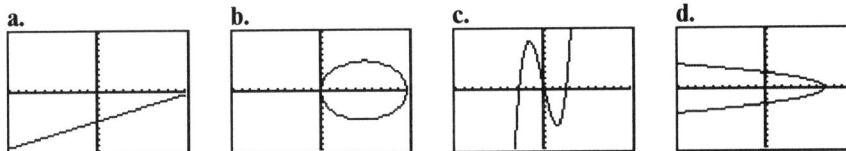

Solution: a.   This graph is the graph of a function because no vertical line will intersect this graph more than once.
b.   This graph is not the graph of a function. Note that vertical lines can be drawn that intersect the graph in two points.
c.   This is the graph of a function.
d.   This is not the graph of a function.

EXERCISES

Exercises 1–6. Find the domain and range of each relation. Also determine whether the relation is a function. See Example 1.

1. $\{(5, 2), (17, 0), (3, -4), (6, 0)\}$ 2. $\{(1, 3), (4, -10), (3, 5), (2, 2)\}$

3. $\{(1, 2), (2, 1), (3, 3), (4, 2), (2, 3)\}$ 4. $\{(2, 2), (1, 1), (3, 2), (4, 1), (0, 4)\}$

5. $\{(a, 1), (b, 1), (c, 1), (2, 4)\}$ 6. $\{(a, a), (b, 3), (c, a), (5, 1), (a, b)\}$

Exercises 7–12. If $f(x) = 5x - 4$, $g(x) = 4 - x^2$, and $h(x) = 2x^2 + 1$, find the following.

7. $f(4)$ **8.** $f(-2)$ **9.** $h(0)$

10. $h(-1)$ **11.** $g(4)$ **12.** $g(-2)$

C *Exercises 13–14. Use a grapher to graph the relation in a standard window. Decide whether the relation is a function. See Example 2.*

13. $y = 3x^{1/3}$ **14.** $y = -1.3x^2 + 9x - 5.5$

Exercises 15–16. Use the vertical line test to determine whether each graph is the graph of a function. See Example 2.

15.

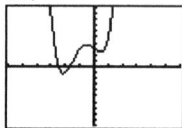

16.

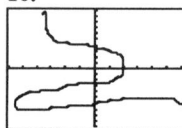

3.3   Graphing Linear Functions

EXAMPLE 1   *Objective 1: Graph linear functions.*

Graph the linear functions $f(x) = \frac{1}{2}x$ and $g(x) = \frac{1}{2}x - 4$ on the same set of axes.

Solution: To graph $f(x)$ and $g(x)$, find ordered pair solutions. The values for $g(x)$ may be obtained by subtracting 4 from the corresponding values for $f(x)$.

x	$f(x) = \frac{1}{2}x$	$g(x) = \frac{1}{2}x - 4$
-2	-1	-5
0	0	-4
2	1	-3
4	2	-2

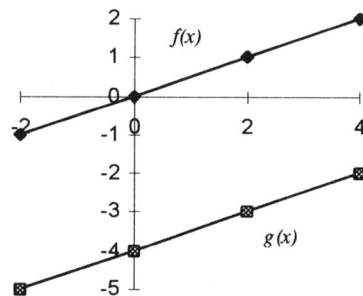

The graph of $g(x)$ is the same as the graph of $f(x)$ shifted down 4 units.

EXAMPLE 2   *Objective 2: Graph linear functions by finding intercepts.*

Graph $4x + y = -3$ by plotting intercept points.

Solution: Let $y = 0$ to find the x-intercept and $x = 0$ to find the y-intercept.

If $y = 0$ then
$$4x + 0 = -3$$
$$4x = -3$$
$$x = -\frac{3}{4}$$

If $x = 0$ then
$$4(0) + y = -3$$
$$0 + y = -3$$
$$y = -3$$

The x-intercept is $-\dfrac{3}{4}$ and the y-intercept is -3. We find a third ordered pair solution to check our work. If we let $y = -7$, then $x = 1$. Plot these three points. The graph of $4x + y = -3$ is the line drawn through these points, as shown.

x	y
$-\dfrac{3}{4}$	0
0	-3
1	-7

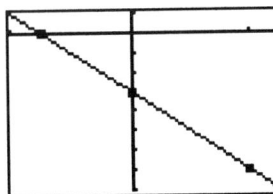

Notice from the graph that the equation $4x + y = -3$ describes a linear function.

EXERCISES

Exercises 1–8. Graph each linear function. See Example 1.

1. $f(x) = 3x$

2. $f(x) = -3x$

3. $f(x) = 3x + 5$

4. $f(x) = -3x - 7$

5. $f(x) = \dfrac{1}{4}x$

6. $f(x) = \dfrac{1}{4}x - 3$

7. $f(x) = \dfrac{1}{4}x + 3$

8. $f(x) = 2x - 5$

Exercises 9–14. Graph each linear function by finding x- and y-intercepts. See Example 2.

9. $x + y = 7$

10. $2x + y = 2$

11. $x + 2y = 5$

12. $2x + 2y = 8$

13. $x - 5y = 10$

14. $x - 3y = -12$

C *Exercises 15–16. Use a grapher to graph each function by first solving for y.*

15. $6x + 9y = -8$

16. $7x - 3y = 12$

Exercises 17–20. Match each equation with its graph.

(a)

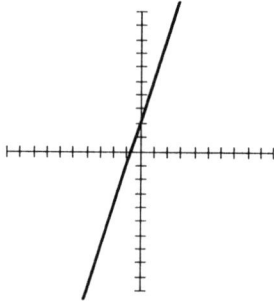

(b)

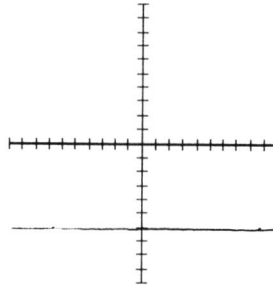

(c)

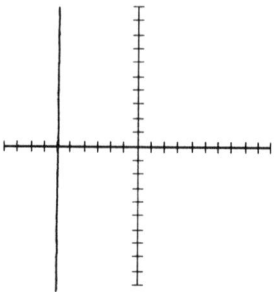

(d)

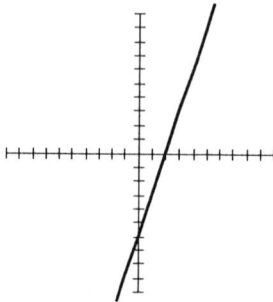

17. $3x - y = 6$ **18.** $3x - y = -2$ **19.** $y = -6$ **20.** $x = -6$

21. Suppose you started the month with $400 in your checking account. You must spend $350 per week on living expenses. You will earn $600 for each 40-hour week that you work. The linear equation that models the balance y in your checking account is $y = 250x + 400$, where x represents the number of weeks you work.

   (a) Complete the ordered pair solution (0,) of this equation. Describe the situation that this solution corresponds to.

   (b) Graph the linear equation.

   (c) What will be the balance of your checking account if you work for 5 weeks?

22. The number of teachers and librarians working in public schools in the United States each year can be estimated by the linear function $f(t) = 52{,}500t + 2{,}865{,}700$, where t is the number of years after 1990 and $f(t)$ is the total number of teachers and librarians working in public schools. (data from National Center for Education Statistics)

   (a) Use this function to approximate the number of teachers and librarians working in public schools in the year 2004. (Hint: Find $f(14)$.)

   (b) Use the given function to predict in what year the number of teachers and librarians exceeds 3,740,000. (Hint: Let $f(t) = 3{,}740{,}000$ and solve for t.)

   (c) Use this function to approximate the number of teachers and librarians working in public schools in the present year.

3.4   The Slope of a Line

EXAMPLE 1   *Objective 1: Find the slope of a line given two points on the line.*

Find the slope of the line containing the points (3, 12) and (–1, 2)

Solution: Use the slope formula. It does not matter which point we call (x_1, y_1) and which point we call (x_2, y_2). Let $(x_1, y_1) = (-1, 2)$ and $(x_2, y_2) = (3, 12)$.

$$m = \frac{y_2 - y_1}{x_2 - x_1}$$

$$= \frac{12 - 2}{3 - (-1)} = \frac{10}{4} = \frac{5}{2}$$

Because the slope is positive, the graph of the line containing these two points moves **upward**, or increases, as we go from left to right. Specifically, the line rise 5 units for every 2-unit move to the right.

EXAMPLE 2   *Objective 2: Find the slope of a line given the equation of a line.*

Find the slope and the y-intercept of the line whose equation is $6x - 2y - 3 = 0$.

Solution: Write the equation in slope-intercept form by solving for *y*.

$6x - 2y - 3 = 0$	
$6x - 2y = 3$	Add 3 to both sides.
$-2y = -6x + 3$	Subtract $6x$ from both sides.
$\dfrac{-2y}{-2} = \dfrac{-6x}{-2} + \dfrac{3}{-2}$	Divide both sides by –2.
$y = 3x - \dfrac{3}{2}$	Simplify.

The coefficient of *x*, 3, is the slope, and the constant term $-\dfrac{3}{2}$ is the *y*-intercept.

EXAMPLE 3   *Objective 3: Compare the slopes of parallel and perpendicular lines.*

Are the following pairs of lines parallel, perpendicular, or neither?

a. $y = -3x + 12; \quad y = -\dfrac{1}{3}x - 4$

b. $y = \dfrac{2}{3}x - 5; \quad y = -\dfrac{3}{2}x + 7$

Solution: The equations are already written in slope-intercept form.

a.   The slope of the first line is –3 and the slope of the second line is $-\dfrac{1}{3}$, so the slopes are not equal, and, thus, the lines are not parallel. The product of these slopes is $(-3)\left(-\dfrac{1}{3}\right) = 1$, not –1 as required to be perpendicular lines. These lines are neither parallel nor perpendicular.

    b. The slope of the first line is $\dfrac{2}{3}$ and the slope of the second line is $-\dfrac{3}{2}$. Again, the slopes are not equal, so the lines are not parallel. The product of the slopes is $\dfrac{2}{3}\left(-\dfrac{3}{2}\right) = -1$, so the lines are perpendicular.

EXERCISES

Exercises 1–6. Find the slope of the line that goes through the given points. See Example 1.

1. $(3, 5), (7, -2)$ **2.** $(6, 18), (0, 0)$ **3.** $(5, 1), (-12, 3)$

4. $(-1, 1), (2, -1)$ **5.** $\left(-\dfrac{1}{2}, \dfrac{3}{4}\right), \left(4, \dfrac{1}{4}\right)$ **6.** $\left(\dfrac{1}{3}, -2\right), \left(\dfrac{2}{3}, 3\right)$

Exercises 7–12. Find the slope and y-intercept of each line. See Example 2.

7. $x + y = 7$ **8.** $2x + y = 2$ **9.** $x + 2y = 5$

10. $2x + 2y = 8$ **11.** $x - 5y = 10$ **12.** $x - 3y = -12$

C *Exercises 13–14. Use a grapher to graph each function and then use the trace feature to complete each ordered pair solution. (Trace as close as you can to the given x- or y-coordinate and approximate the other unknown coordinate to one decimal place.)*

13. $y = -\dfrac{4}{3}x + 6$ **14.** $y = 2x + \dfrac{3}{5}$
      $x = 4, \ y = ?$ $x = ?, \ y = -3$

Exercises 15–18. Are the following pairs of lines parallel, perpendicular, or neither? See Example 3.

15. $y = 7x - 8$ and $y = -7x + 3$ **16.** $y = -\dfrac{6}{7}x + 2$ and $y = \dfrac{7}{6}x - 3$

17. $3x - y = 6$ and $2x + 6y = 13$ **18.** $2x - 5y = 7$ and $-x + \dfrac{5}{2}y = -4$

19. The number of poodles registered with the American Kennel Club in 1992 was 73,449. In 1994 there were 61,775 poodles registered. Assuming that the trend in American Kennel Club poodle registration is linear, find the slope of the line described by these points. (data from American Kennel Club)

20. The percent of households in the United States that used a black and white television set in 1980 was 51 percent. By 1993, black and white television sets were used in only 20 percent of households in the United States. The change in black and white television usage over this period was linear. Find the slope of the line representing the change black and white television usage. (data from U.S. Department of Energy)

3.5   Equations of Lines

EXAMPLE 1   *Objectives 1 and 2: Use the slope-intercept form to find the equation of a line and graph a line given its slope and y-intercept.*

Write an equation of the line with y-intercept 2 and slope of $-\dfrac{5}{2}$. Graph the line.

Solution: Let $m = -\dfrac{5}{2}$ and $b = 2$, and write the equation in slope-intercept form, $y = mx + b$.

$$y = mx + b$$

$$y = -\frac{5}{2}x + 2 \qquad \text{Let } m = -\frac{5}{2} \text{ and } b = 2.$$

To graph the line, we need two points. One point is the y-intercept $(0, 2)$, and we will use the slope $m = -\dfrac{5}{2} = \dfrac{-5}{2}$ to find another point. Start at $(0, 2)$ and move vertically down 5 units because the numerator of the slope is -5; then move horizontally 2 units to the right because the denominator of the slope is 2. We arrive at the point $(2, -3)$. The line through $(0, 2)$ and $(2, -3)$ has the required slope of $-\dfrac{5}{2}$.

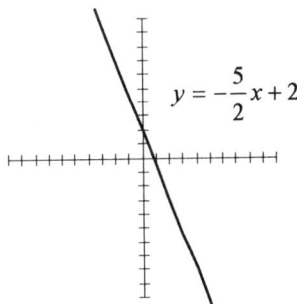

$$y = -\frac{5}{2}x + 2$$

EXAMPLE 2   *Objective 3: Use the point-slope form to find the equation of a line.*

Find an equation of the line through the points (3, 12) and (−1, 2). Write the equation in standard form.

Solution: First, find the slope of the line.

$$m = \frac{12-2}{3-(-1)} = \frac{10}{4} = \frac{5}{2}$$

Next, make use of the point-slope form.

$$y - y_1 = m(x - x_1)$$

$y - 12 = \dfrac{5}{2}(x-3)$	Let $m = \dfrac{5}{2}$ and $(x_1, y_1) = (3, 12)$.
$2y - 24 = 5(x-3)$	Multiply both sides by 2.
$2y - 24 = 5x - 15$	Remove parentheses.
$2y = 5x + 9$	Add 24 to both sides.
$-5x + 2y = 9$	Subtract $5x$ from both sides.

EXERCISES

Exercises 1–4. Find an equation of each line with the given slope and y-intercept. Graph each equation. See Example 1.

1. Slope: 2; *y*-intercept: 0

2. Slope: –3; *y*-intercept: 2

3. Slope: $\dfrac{1}{2}$; *y*-intercept: –3

4. Slope: $-\dfrac{4}{3}$; *y*-intercept: –1

Exercises 5–7. Find an equation of the line passing through the given points. Write the equation in standard form. See Example 2.

5. (3, 5), (7, –2)

6. $\left(-\dfrac{1}{2}, \dfrac{3}{4}\right), \left(4, \dfrac{1}{4}\right)$

7. (5, 1), (–12, 3)

Exercises 8–10. Find an equation of the line passing through the given points. Write the equation using function notation.

8. (–1, 1), (2, –1)

9. (6, 18), (0, 0)

10. $\left(\dfrac{1}{3}, -2\right), \left(\dfrac{2}{3}, 3\right)$

Exercises 11–14. Find an equation of each line. Write the equation using function notation.

11. Through (1, 1) and parallel to $x + y = 7$

12. Through (–3, 6) and parallel to $x - 5y = 10$

13. Through $(-1, -4)$ and perpendicular to $x + 2y = 5$ **14.** Through $(2, -5)$ and perpendicular to $x - 3y = -12$

15. The number of Siberian huskies registered with the American Kennel Club in 1992 was 26,057. In 1994 there were 24,084 Siberian huskies registered. Assuming that the trend in American Kennel Club poodle registration is linear, write an equation of the line described by these points. Use ordered pairs of the form (years past 1990, number of registered Siberian huskies). (data from American Kennel Club)

16. The percent of households in the United States that used an outdoor gas grill in 1980 was 9 percent. By 1993, outdoor grill usage had increased to 29 percent of households in the United States. The change in gas grill usage over this period was linear. Write an equation of the line that describes the change in outdoor gas grill usage. Use ordered pairs of the form (years past 1980, percent of households using outdoor gas grills). (data from U.S. Department of Energy)

3.6   Graphing Linear Inequalities

EXAMPLE 1   *Objective 1: Graph linear inequalities.*

Graph $-2x + 3y \geq 9$.

Solution: First graph the boundary line $-2x + 3y = 9$ as a solid line because the inequality symbol is $\geq$.

Test a point not on the boundary line to determine which half-plane contains points that satisfy the inequality. From the graph, you can see that the point $(0, 0)$ lies below the boundary line; choose this point as the test point.

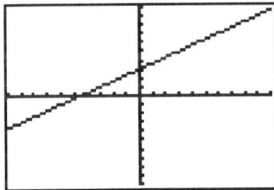

$$-2x + 3y \geq 9$$
$$-2(0) + 3(0) \geq 9 \qquad \text{Let } x = 0 \text{ and } y = 0.$$
$$0 + 0 \geq 9 \qquad \text{False.}$$

This point does not satisfy the inequality, so the correct half-plane is above the boundary line, not below it. The graph of $-2x + 3y \geq 9$ is the boundary line together with the shaded region shown next.

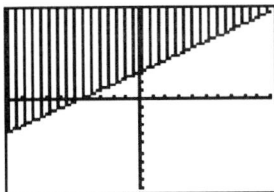

EXAMPLE 2   *Objective 2: Graph the intersection or union of two linear inequalities.*

Graph the intersection of $y \leq x + 5$ and $y \geq -2x - 4$.

Solution: Graph each inequality. The intersection of the two graphs is all points common to both regions, as shown by the heaviest shading in the third graph.

$y \le x + 5$

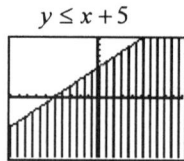

$y \ge -2x - 4$

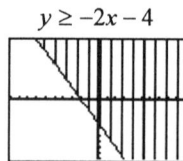

$y \le x + 5$ and $y \ge -2x - 4$

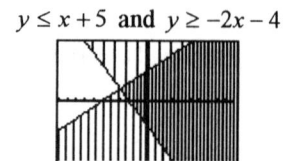

EXERCISES

Exercises 1–4. Graph each inequality. See Example 1.

1. $2x + 2y > -3$

2. $-x + 5y \le 4$

3. $x - 3y \ge 12$

4. $2x + y \ge -5$

Exercises 5–8. Graph each union or intersection. See Example 2.

5. The intersection $y \le -2x + 5$ and $y \ge \dfrac{1}{2}x - 6$

6. The intersection $y \le 4x - 3$ and $y \ge 4x - 3$

7. The union $2x + 3y \le 9$ or $4x + 6y \le 0$

8. The union $x + y \le 7$ or $x - y \le 8$

Exercises 9–12. Match each inequality with its graph.

(a)

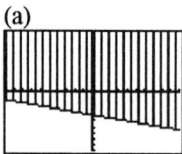

(b)

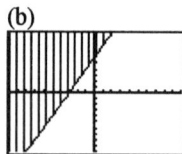

(c)

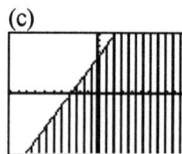

(d)

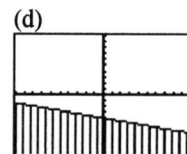

9. $y \le -\dfrac{1}{4}x - 4$

10. $y \ge -\dfrac{1}{4}x - 4$

11. $y \le 2x + 6$

12. $y \ge 2x + 6$

13. You have two hobbies: fishing and baking bread. During your vacation, you decide to spend at least 12 hours per week baking bread and no more than 25 hours per week fishing. Let x represent the hours spent baking and y represent the hours spent fishing. Write two inequalities that model this situation and graph their intersection. Interpret the meaning of the intersection of these two inequalities.

Chapter 3　Hints and Warnings

- To **plot or graph an ordered pair** (x, y), start at the origin. If x is positive, move x units to the right along the x-axis. If x is negative, move x units to the left along the x-axis. From that point, if y is positive, move y units upward parallel to the y-axis; or if y is negative, move y units downward parallel to the y-axis.
- To **test whether the graph of a relation is a function,** examine the graph to see if any vertical line can be drawn that will intersect the graph more than once. If not, the graph is the graph of a function.
- To **graph a linear function,** find three ordered pair solutions. Graph the solutions and draw a line through the plotted points. *Don't forget that you can also use the y-intercept (or any other point on the line) and the slope of the line to draw the line.*
- To **find an** x-**intercept,** let $y = 0$ or $f(x) = 0$ and solve for x.
- To **find a** y-**intercept,** let $x = 0$ and solve for y.
- To **find the equation of a line** given the slope and a point on the line, use the point-slope form of the equation of a line $y - y_1 = m(x - x_1)$ by substituting the value of the slope for m and the coordinates of the point for x_1 and y_1.
- To **graph a linear inequality:**
  1) Graph the boundary line by graphing the related equation. Draw a solid line if the inequality symbol is $\leq$ or $\geq$. Draw a dashed line if the inequality symbol is $<$ or $>$.
  2) Choose a test point not on the line. Substitute its coordinates into the original inequality.
  3) If the resulting inequality is true, shade the half-plane that contains the test point. If the inequality is not true, shade the half-plane that does not contain the test point.

CHAPTER 3　Practice Test

1. Name the quadrant in which each point is located: $(-2, 8)$, $(3, 11)$, $(4, -5)$, and $(-12, -1)$.

2. Complete the ordered pair solution (, -4) of the equation $-x + 5y = 4$.

Graph each line.

3. $x - 3y = 12$

4. $2x + y = -5$

5. $x + y - 4 = 0$

6. $y = -7$

7. Find the slope of the line that passes through the points $(1, -8)$ and $(5, 10)$.

8. Find the slope and the y-intercept of the line $-8x + 2y = -6$.

Graph each function. Suggested x-values have been given for ordered pair solutions.

9. $f(x) = -x^2 + 4$
   Let $x = -3, -2, -1, 0, 1, 2, 3$

10. $g(x) = |2x - 4| - 3$
    Let $x = -1, 0, 1, 2, 3, 4, 5$

Find an equation of each line satisfying the conditions given.

11. Horizontal line through $(-3, 3)$

12. Vertical line through $(5, 7)$

13. Perpendicular to $y = -4$; through $(-5, -9)$

14. Parallel to $y = \dfrac{3}{4}$; through $(6, -1)$

15. Through $(1, 4)$; slope -2

16. Through $(10, -8)$; slope 4

17. Through $(3, 3)$; parallel to $x + y = -4$

18. Through $(-2, 5)$; perpendicular to $x + 2y = 10$

Graph each inequality.

19. $2x + 2y \leq 9$

20. $5x - y \geq -3$

Find the domain and range of each relation. Also determine whether the relation is also a function.

21. $\{(0, 7), (3, -3), (-2, 4), (3, 11), (1, 4)\}$

22. $\{(2, -6), (1, 12), (3, 4), (-1, 12), (-3, 4)\}$

23. The total number of commercial radio stations in the United States each year is given by the linear function $f(x) = 151x + 9429$, where x is the number of years past 1990. (Source: M Street Corporation)
   (a) Find the number of commercial radio stations in 1995.
   (b) Predict the number of commercial radio stations in 2002.
   (c) Predict the first year that the number of commercial radio stations is greater than 10,750.

CHAPTER 4

Systems of Equations

4.1 Solving Systems of Linear Equations in Two Variables

EXAMPLE 1   *Objective 1: Solve a system by graphing.*

Solve the system by graphing. Estimate the solution.

$$\begin{cases} 2x - y = 6 \\ 7x + 2y = 10 \end{cases}$$

Solution:

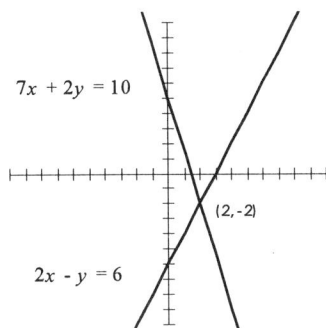

These lines intersect in one point as shown. The coordinates of the point of intersection appear to be $(2, -2)$. Check this estimated solution by replacing x with 2 and y with -2 in **both** equations.

$2x - y = 6$	First equation.	$7x + 2y = 10$	Second equation.
$2(2) - (-2) = 6$	Let $x = 2$ and $y = -2$.	$7(2) + 2(-2) = 10$	Let $x = 2$ and $y = -2$.
$4 + 2 = 6$	Simplify.	$14 - 4 = 10$	Simplify.
$6 = 6$	True.	$10 = 10$	True.

The ordered pair $(2, -2)$ does satisfy both equations. We conclude that $(2, -2)$ is the solution of the system.

EXAMPLE 2   *Objective 2: Solve a system by substitution.*

Use substitution to solve the system

$$\begin{cases} 3x - 5y = -7 & \text{First equation.} \\ 2x + \dfrac{y}{3} = 8 & \text{Second equation.} \end{cases}$$

Solution: First, multiply the second equation by its least common denominator in order to write this system without fractions. We multiply the second equation by 3.

$$\begin{cases} 3x - 5y = -7 & \text{First equation.} \\ 6x + y = 24 & \text{Second equation.} \end{cases}$$

We now solve the second equation for y.

$$6x + y = 24 \qquad \text{Second equation.}$$
$$y = 24 - 6x \qquad \text{Solve for } y.$$

45

Next, replace y with $24 - 6x$ in the first equation.

$$3x - 5y = -7 \qquad \text{First equation.}$$

$$3x - 5(24 - 6x) = -7$$

$$3x - 120 + 30x = -7$$

$$33x = 113$$

$$x = \frac{113}{33} \qquad \text{Solve for } x.$$

The x-coordinate is $\dfrac{113}{33}$. To find the y-coordinate, replace x with $\dfrac{113}{33}$ in the equation $y = 24 - 6x$. Then

$$y = 24 - 6x = 24 - 6\left(\frac{113}{33}\right) = \frac{264}{11} - \frac{226}{11} = \frac{38}{11}$$

The solution set is $\left\{\left(\dfrac{113}{33}, \dfrac{38}{11}\right)\right\}$.

EXAMPLE 3   *Objective 3: Solve a system by elimination.*

Use the elimination method to solve the system

$$\begin{cases} 12x - 5y = 33 \\ 4x - 3y = 23 \end{cases}$$

Solution: To eliminate x when the equations are added, multiply both sides of the second equation by -3.

$$\begin{cases} 12x - 5y = 33 \\ -3(4x - 3y) = -3(23) \end{cases} \quad \text{simplifies to} \quad \begin{cases} 12x - 5y = 33 \\ -12x + 9y = -69 \end{cases}$$

Next, add the left and right sides.

$$\begin{array}{rcrcr} 12x & - & 5y & = & 33 \\ -12x & + & 9y & = & -69 \\ \hline & & 4y & = & -36 \\ & & y & = & -9 \end{array}$$

To find x, let $y = -9$ in either equation of the system.

$$4x - 3y = 23$$

$$4x - 3(-9) = 23$$

$$4x + 27 = 23$$

$$4x = -4$$

$$x = -1$$

The solution set is $\{(-1, -9)\}$.

EXERCISES

Exercises 1–3. Solve each system by graphing. See Example 1.

1. $\begin{cases} x - y = 0 \\ x + y = 8 \end{cases}$

2. $\begin{cases} -8x + 3y = -6 \\ -4x + 3y = 6 \end{cases}$

3. $\begin{cases} \dfrac{x}{3} + y = -5 \\ -5x + y = 11 \end{cases}$

Exercises 4–6. Solve each system by the substitution method. See Example 2.

4. $\begin{cases} 5x + 6y = 40 \\ 2x - 3y = -8 \end{cases}$

5. $\begin{cases} x - 2y = 8 \\ 3x + 4y = 16 \end{cases}$

6. $\begin{cases} 10x + y = 3 \\ -2x + 5y = -12 \end{cases}$

Exercises 7–9. Solve each system by the elimination method. See Example 3.

7. $\begin{cases} 3x - 2y = 9 \\ 5x + 2y = 7 \end{cases}$

8. $\begin{cases} 14x + 4y = 5 \\ 7x - 12y = 60 \end{cases}$

9. $\begin{cases} -2x - 5y = 11 \\ 3x - 3y = 17 \end{cases}$

C *Exercises 10–13. Use a grapher to solve each system of equations. Approximate the solutions to two decimal places.*

10. $\begin{cases} y = 4.5x - 18 \\ y = -8.3x + 22 \end{cases}$

11. $\begin{cases} y = 6.75x + 5.25 \\ y = -7x - 13.75 \end{cases}$

12. $\begin{cases} 2.2x - 3.7y = 40 \\ 3.4x + 12y = 19.2 \end{cases}$

13. $\begin{cases} -14x + 7.5y = 23 \\ 3x - 4.9y = 12.1 \end{cases}$

Exercises 14–19. Solve each system of equations.

14. $\begin{cases} 2x - y = 31 \\ 3x + 2y = -45 \end{cases}$

15. $\begin{cases} x + 13y = 26 \\ -8x + y = -15 \end{cases}$

16. $\begin{cases} 4x - \dfrac{y}{2} = 7 \\ -3x + 10y = 14 \end{cases}$

17. $\begin{cases} x + 2y = 20 \\ 2x - \dfrac{y}{3} = 10 \end{cases}$

18. $\begin{cases} \dfrac{x}{4} + \dfrac{3x}{7} = -\dfrac{5}{2} \\ -\dfrac{x}{2} + \dfrac{y}{5} = 6 \end{cases}$

19. $\begin{cases} \dfrac{x}{12} - \dfrac{5y}{3} = \dfrac{7}{4} \\ 6x + 120y = -126 \end{cases}$

20. A fruit farm raises and sells fresh strawberries. The revenue equation for strawberries is $y = 2.3x$, where x is the number of quarts of strawberries sold and y is the revenue for selling x quarts. The cost equation for strawberries is $y = 0.95x + 2000$, where x is the number of quarts of strawberries sold and y is the total cost for producing x quarts. Use these equations to find the number of quarts of strawberries that must be sold for the fruit farm to break even.

21. The revenue equation for a hair salon is $y = 16x$, where y is the income from giving x haircuts. The salon's cost equation for a month is $y = 4.25x + 7500$, where y is the total cost of giving x haircuts. How many haircuts must the salon give during a month to break even for that month?

22. Examine the graph given at the right. Suppose the lines represent a cost equation and a revenue equation for a small company.
  (a) Which line is the cost equation? Which line is the revenue equation? Explain your reasoning.
  (b) Estimate the break even point from the graph (suppose each tick mark on the each axis represents 100 units)

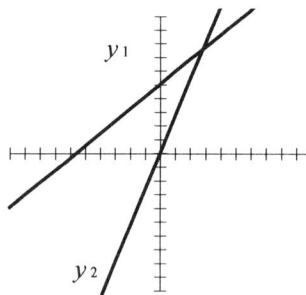

4.2   Solving Systems of Linear Equations in Three Variables

EXAMPLE 1   *Objective 2: Solve a system of three linear equations in three variables.*

Solve the system:
$$\begin{cases} x + y + z = 3 & (1) \\ 2x + y - z = 4 & (2) \\ x + 2z = -4 & (3) \end{cases}$$

Solution: If equations (1) and (2) are added, z is eliminated, and the new equation is
$$3x + 2y = 7 \qquad (4)$$

To eliminate z again, multiply both sides of equation (2) by 2, and add the resulting equation to equation (3). Then

$$\begin{cases} 2(2x + y - z) = 2(4) \\ x + 2z = -4 \end{cases} \quad \text{simplifies to} \quad \begin{array}{r} 4x + 2y - 2z = 8 \\ x + 2z = -4 \\ \hline 5x + 2y = 4 \qquad (5) \end{array}$$

Next, solve for x and y using equations (4) and (5). Multiply both sides of equation (4) by -1, and add the resulting equation to (5).

$$\begin{cases} -1(3x + 2y) = -1(7) \\ 5x + 2y = 4 \end{cases} \quad \text{simplifies to} \quad \begin{array}{r} -3x - 2y = -7 \\ 5x + 2y = 4 \\ \hline 2x = -3 \end{array}$$

This tells us that $x = -\dfrac{3}{2}$. Replace x with $-\dfrac{3}{2}$ in equation (4) or (5).

$$3x + 2y = 7 \qquad \text{Equation (4)}$$

$$3\left(-\frac{3}{2}\right) + 2y = 7 \qquad \text{Let } x = -\frac{3}{2}.$$

$$-\frac{9}{2} + 2y = 7$$

$$2y = \frac{23}{2}$$

$$y = \frac{23}{4}$$

Finally, replace x with $-\frac{3}{2}$ and y with $\frac{23}{4}$ in equation (1), (2), or (3) and solve for z.

$$x + y + z = 3 \qquad \text{Equation (1)}$$

$$-\frac{3}{2} + \frac{23}{4} + z = 3$$

$$z = 3 + \frac{3}{2} - \frac{23}{4}$$

$$z = \frac{12}{4} + \frac{6}{4} - \frac{23}{4} = -\frac{5}{4}$$

The solution set is $\left\{\left(-\frac{3}{2}, \frac{23}{4}, -\frac{5}{4}\right)\right\}$. Check to see that this satisfies all three equations.

EXERCISES

Exercises 1–6. Solve.

1. $\begin{cases} x + z = -3 \\ 5y = 15 \\ x + y - z = 4 \end{cases}$

2. $\begin{cases} 2x = 26 \\ y - 3z = 21 \\ x + 2y + 5z = 16 \end{cases}$

3. $\begin{cases} x + y + \phantom{\frac{1}{2}}z = 0 \\ -2y + \frac{1}{2}z = 4 \\ -x + y \phantom{+\frac{1}{2}z} = 12 \end{cases}$

4. $\begin{cases} 4y + z = 10 \\ x + 2z = 5 \\ -3x + y = -8 \end{cases}$

5. $\begin{cases} x - y + 2z = 0 \\ x + 2y + z = 0 \\ 2x - y - z = 0 \end{cases}$

6. $\begin{cases} -2x + y - 2z = -6 \\ x + y + z = 1 \\ x - y - z = -4 \end{cases}$

7. Write a system of three linear equations in three variables that has (0, 0, 0) as a solution. (There are many possibilities.)

8. Write a system of three linear equations in three variables that has (1, −10, 5) as a solution. (There are many possibilities.)

4.3   Systems of Linear Equations and Problem Solving

EXAMPLE 1   *Objective 1: Solve problems that can be modeled by a system of linear equations.*

A bullet train in France leaves Paris heading toward Bordeaux at the same time that another bullet train leaves Bordeaux heading toward Paris. Bordeaux is approximately 307 miles from Paris by train. After 1 hour and 45 minutes, the two trains pass one another. If the train leaving Bordeaux travels 15 mph slower than the other train, how far from Paris did the two trains meet?

Solution: 1.   UNDERSTAND. Read and reread the problem. Guess that the speed of the train leaving Bordeaux is 60 mph. That means that the other train is traveling at a rate of 75 mph because we are told that the train leaving Bordeaux travels 15 mph slower than the other train. To find the distance from Paris that the trains pass, we must find the distance traveled by each. The train leaving Paris travels a distance of rate • time = $75(1.75) = 131.25$ miles. The other train travels a distance of rate • time = $60(1.75) = 105$ miles. The sum of these distances is 236.25 miles, not the required distance of 307 miles between the two cities. Our guess is incorrect, but we now have a better idea of how to model the problem.

2.   ASSIGN. Let x = the speed of the train leaving Paris
             y = the speed of the train leaving Bordeaux

3.   ILLUSTRATE. We summarize the information on the following chart. The trains each traveled 1 hour and 45 minutes, or 1.75 hours. Use the fact that distance = rate • time to fill in the chart.

	Rate	•	Time	=	Distance
Train leaving Paris	x		1.75		$1.75x$
Train leaving Bordeaux	y		1.75		$1.75y$

4.   TRANSLATE. We translate into two equations.

In words:	Distance traveled by train leaving Paris	added to	Distance traveled by train leaving Bordeaux	is	307 miles

Translate: $1.75x$ + $1.75y$ =   307

In words:	Train leaving Bordeaux	is	15 mph slower than the other

Translate: y = $x - 15$

5.   COMPLETE. Here we solve the system
$$\begin{cases} 1.75x + 1.75y = 307 \\ y = x - 15 \end{cases}$$

The substitution method is appropriate. Replace y with $x - 15$ in the first equation and solve for x.

$$1.75x + 1.75y = 307$$
$$1.75x + 1.75(x - 15) = 307 \qquad \text{Let } y = x - 15.$$
$$1.75x + 1.75x - 26.25 = 307$$
$$3.5x = 333.25$$
$$x \approx 95.2$$

To find y, replace x with 95.2 in the equation $y = x - 15$. Then $y = 95.2 - 15 = 80.2$.

6. INTERPRET. We have found the speeds of both trains, but the problem asked for the number of miles from Paris that the two trains pass. This can be found by finding the miles traveled by the train leaving Paris; that is, evaluating $1.75x$ for $x = 95.2$. This is 166.6 miles. *Check:* If the train leaving Paris travels at a rate of 95.2 mph and the train leaving Bordeaux travels at a rate of 80.2 miles,

$$95.2(1.75) + 80.2(1.75) = 166.6 + 140.35$$

≈ 307 miles, the given distance between the cities

State: The two trains meet at a distance of 166.6 miles from Paris.

EXERCISES

Exercises 1–2. Solve. See Example 1.

1. A bicyclist leaves Morris, Minnesota, and travels toward Breckenridge, Minnesota. At the same time, a car leaves Breckenridge traveling toward Morris. Morris is 50 miles from Breckenridge. After 37.5 minutes, the car passes the bicyclist. If the car is traveling 3 times faster than the bicyclist, how far from Morris did the car and bicycle pass?

2. A truck leaves Lubbock, Texas, and travels toward Del Rio, Texas, which is 345 miles away. At the same time, a second truck leaves Del Rio traveling toward Lubbock on the same path as the first truck. The trucks pass after 3 hours. If the truck leaving Del Rio is traveling 7 mph slower than the other truck, how far from Lubbock did the trucks pass one another?

Exercises 3–18. Solve.

3. One number is six less than a second number. Half the first number is 25 more than 4 times the second number. Find the two numbers.

4. The sum of two numbers is 95. One-half of one number is twice the second number. Find the two numbers.

5. An Indy car races with the wind down the front stretch of an oval racetrack at a speed of 185 mph. Along the back stretch, the same car drives at a speed of 171 mph against the wind. Find the speed of the Indy car in still air and the speed of the wind.

6. A rowing team can row 16 kilometers in 1 hour upstream and 20.2 kilometers in 1 hour downstream. Find how fast the team can row in still water, and find the speed of the current.

7. A dozen apples and a half dozen oranges cost $6.90. A half dozen apples and a dozen oranges cost $7.50. Find the price of a single orange and a single apple.

8. Eight flats of impatiens and 2 flats of marigolds cost $15.50. Three flats of impatiens and 5 flats of marigolds cost $10.02. Find the price per flat of each kind of flower.

9. The perimeter of a pentagon (five-sided figure) is 210 centimeters. The shortest side is half as long as the longest side. The other three sides are equally long and each are 5 inches shorter than the longest side. Find the length of all five sides.

10. The perimeter of a triangle is 68 inches. If two sides are equally long and the third side is 14 inches longer than the other two, find the lengths of the three sides.

11. One truck rental company charges a $50 rental fee plus $0.30 per mile. A second truck rental company charges a $40 rental fee and $0.45 per mile. Find the number of miles for which the total cost of renting a truck is the same at both rental companies.

12. A party goods store rents helium tanks and charges $2 per cubic foot of helium used plus a $10 processing charge. A second store charges $2.20 per cubic foot of helium used plus a $7 processing charge. Find the number of cubic feet of helium for which the total rental cost is the same at both stores.

13. A craftsperson makes and sells wooden plant stands. She sells each stand for $38. The materials to make each stand cost $13, and she bought $300 worth of equipment to make the stands. What is her break even point?

14. A florist stocks and sells fresh-cut roses. He pays $10 per dozen wholesale for the roses, and sells the roses for $19.95 per dozen. If it costs $400 per year to run the refrigeration unit for storing the roses, what is his break even point?

15. The sum of three numbers is 100. The third number is three times the first number, and the second number is 12 less than twice the third. Find the three numbers.

16. The sum of three numbers is 0. The difference between twice the largest and the smallest is 50. The middle number is half the sum of the smallest and largest. Find the numbers.

17. Data (x, y) for a company's sales over time are $(1, 45)$, $(7, 171)$, and $(13, 153)$, where x represents time in months and y represents sales in thousands. Find the values of a, b, and c such that the equation $y = ax^2 + bx + c$ models this data. To do so, substitute each ordered pair into the equation. Each time, the result is an equation in three unknowns: a, b, and c. Then solve the resulting system of three linear equations.

18. Monthly normal rainfall data (x, y) for Juneau, Alaska, are $(2, 3.7)$, $(4, 2.8)$, $(9, 6.7)$, where x represents time in months (with $x = 2$ representing February) and y represents rainfall in inches. Find the values of a, b, and c such that the equation $y = ax^2 + bx + c$ models this data. (data from National Climatic Data Center)

C *Exercises 19–20. Use a grapher with the given cost function C(x) and revenue function R(x) to estimate (accurate to one whole unit) the number of units x that must be sold to break even.*

19. $C(x) = 7.80x + 1525$
$R(x) = 24.95x$

20. $C(x) = 45.45x + 5500$
$R(x) = 84.99x$

4.4   Solving Systems of Equations by Matrices

EXAMPLE 1   *Objective 1: Use matrices to solve a system of two equations.*

Solve the system using matrices.

$$\begin{cases} -2x + 2y = 13 \\ \quad x - 5y = -2 \end{cases}$$

Solution: The corresponding augmented matrix is $\begin{bmatrix} -2 & 2 & | & 13 \\ 1 & -5 & | & -2 \end{bmatrix}$. Use elementary row operations to write

an equivalent matrix that has 1's along the main diagonal and 0's below each 1 in the main diagonal. To obtain a 1 as the first element in the first row, we can exchange the two rows.

$$\begin{bmatrix} 1 & -5 & | & -2 \\ -2 & 2 & | & 13 \end{bmatrix}$$

Now, write an equivalent matrix with a 0 in the first column below the 1. To do this, multiply row 1 by 2 and add to row 2.

$$\begin{bmatrix} 1 & -5 & | & -2 \\ 2(1)+(-2) & 2(-5)+2 & | & 2(-2)+13 \end{bmatrix} \quad \text{simplifies to} \quad \begin{bmatrix} 1 & -5 & | & -2 \\ 0 & -8 & | & 9 \end{bmatrix}$$

Now continue down the main diagonal and change the –8 to a 1 by use of an elementary row operation. Divide row 2 by –8. Then

$$\begin{bmatrix} 1 & -5 & | & -2 \\ \dfrac{0}{-8} & \dfrac{-8}{-8} & | & \dfrac{9}{-8} \end{bmatrix} \quad \text{simplifies to} \quad \begin{bmatrix} 1 & -5 & | & -2 \\ 0 & 1 & | & -1.125 \end{bmatrix}$$

This last matrix corresponds to the system

$$\begin{cases} x - 5y = -2 \\ \quad y = -1.125 \end{cases}$$

To find x, let $y = -1.125$ in the first equation.

$$\begin{aligned} x - 5y &= -2 && \text{First equation.} \\ x - 5(-1.125) &= -2 && \text{Let } y = -1.125. \\ x + 5.625 &= -2 \\ x &= -7.625 \end{aligned}$$

The solution set is $\{(-7.625, -1.125)\}$. Check to see that this ordered pair satisfies both equations.

EXAMPLE 2   *Objective 2: Use matrices to solve a system of three equations.*

Solve the system using matrices.

$$\begin{cases} x - 3y + z = -1 \\ -2x + 6y - 2z = 3 \\ 2x + y - 2z = 0 \end{cases}$$

Solution: The corresponding augmented matrix is $\left[\begin{array}{ccc|c} 1 & -3 & 1 & -1 \\ -2 & 6 & -2 & 3 \\ 2 & 1 & -2 & 0 \end{array}\right]$. To get a 0 under the 1 in the

first column, multiply the elements of row 1 by 2 and add the new elements to the elements of row 2.

$$\left[\begin{array}{ccc|c} 1 & -3 & 1 & -1 \\ 2(1)-2 & 2(-3)+6 & 2(1)-2 & 2(-1)+3 \\ 2 & 1 & -2 & 0 \end{array}\right] \quad \text{simplifies to} \quad \left[\begin{array}{ccc|c} 1 & -3 & 1 & -1 \\ 0 & 0 & 0 & 1 \\ 2 & 1 & -2 & 0 \end{array}\right]$$

The new second row corresponds to the equation $0 = 1$, which is a false statement for all values of x, y, or z. Hence, this system is inconsistent and has no solution.

EXERCISES

Exercises 1–4. Solve each system of linear equations using matrices. See Example 1.

1. $\begin{cases} x + y = -10 \\ x - 6y = 18 \end{cases}$ **2.** $\begin{cases} x - 3y = 9 \\ x + y = 18 \end{cases}$ **3.** $\begin{cases} x - 5y = 15 \\ 2x + 5y = 12 \end{cases}$ **4.** $\begin{cases} x + 4y = -10 \\ 3x - y = -4 \end{cases}$

Exercises 5–6. Solve, if possible, each system of linear equations using matrices. See Example 2.

5. $\begin{cases} 3x - 4y = 20 \\ -9x + 12y = 11 \end{cases}$ **6.** $\begin{cases} -x + 2y + 6z = -7 \\ \dfrac{1}{2}x - y - 3z = 12 \\ 2x + \dfrac{3}{2}y - z = 0 \end{cases}$

Exercises 7–14. Solve using matrices.

7. $\begin{cases} 3x + 6y = 51 \\ -2x + y = 8 \end{cases}$ **8.** $\begin{cases} -x + \dfrac{1}{2}y = 4 \\ 2x + y = -3 \end{cases}$ **9.** $\begin{cases} -2x - 2y = 5 \\ x + 4y = -20 \end{cases}$

10. $\begin{cases} x + y + z = 0 \\ 2x + y = 10 \\ z = 2 \end{cases}$

11. $\begin{cases} 2x + z = -8 \\ x + y = 7 \\ y + 3z = -10 \end{cases}$

12. $\begin{cases} x - y - z = 9 \\ x + y + z = 32 \\ 2x - z = 0 \end{cases}$

13. $\begin{cases} -x + y + z = -5 \\ -x + 2z = 4 \\ x + 2y + 3z = 12 \end{cases}$

14. $\begin{cases} \dfrac{1}{2}x + \dfrac{1}{4}y - z = 2 \\ x + y + z = 17 \\ 2x + y - 3z = 0 \end{cases}$

C *Exercises 15–16. Solve each system of equations using matrices. Use a grapher to check your solution graphically.*

15. $\begin{cases} y = -\dfrac{3}{4}x + 6 \\ y = 2x + 7 \end{cases}$

16. $\begin{cases} 2.2x + 3.7y = 0 \\ 4x - 1.5y = 10 \end{cases}$

4.5   Solving Systems of Equations by Determinants

EXAMPLE 1   *Objective 2: Use Cramer's rule to solve a system of two linear equations in two variables.*

Use Cramer's rule to solve the system.
$$\begin{cases} -2x + 2y = 13 \\ x - 5y = -2 \end{cases}$$

Solution: Find D, D_x, D_y.

$$D = \begin{vmatrix} -2 & 2 \\ 1 & -5 \end{vmatrix} = (-2)(-5) - (1)(2) = 10 - 2 = 8$$

$$D_x = \begin{vmatrix} 13 & 2 \\ -2 & -5 \end{vmatrix} = (13)(-5) - (-2)(2) = -65 + 4 = -61$$

$$D_y = \begin{vmatrix} -2 & 13 \\ 1 & -2 \end{vmatrix} = (-2)(-2) - (1)(13) = 4 - 13 = -9$$

$$x = \frac{D_x}{D} = \frac{-61}{8} = -7.625 \quad \text{and} \quad y = \frac{D_y}{D} = \frac{-9}{8} = -1.125$$

The solution set is $\{(-7.625, -1.125)\}$. Notice that this result agrees with the result obtained for the same system in Example 1 in Section 4.4 of this Study Guide.

EXAMPLE 2   *Objective 4: Use Cramer's rule to solve a system of three linear equations in three variables.*

Use Cramer's rule to solve the system.
$$\begin{cases} x + y + 3z = 14 \\ 2x = 25 \\ 3y + z = 8 \end{cases}$$

Solution: First, find D, D_x, D_y, and D_z. Beginning with D, we expand by minors of the second row.

$$D = \begin{vmatrix} 1 & 1 & 3 \\ 2 & 0 & 0 \\ 0 & 3 & 1 \end{vmatrix} = -2 \cdot \begin{vmatrix} 1 & 3 \\ 3 & 1 \end{vmatrix} + 0 \cdot \begin{vmatrix} 1 & 3 \\ 0 & 1 \end{vmatrix} - 0 \cdot \begin{vmatrix} 1 & 1 \\ 0 & 3 \end{vmatrix} = -2 \cdot [(1)(1) - (3)(3)] + 0 - 0 = 16$$

$$D_x = \begin{vmatrix} 14 & 1 & 3 \\ 25 & 0 & 0 \\ 8 & 3 & 1 \end{vmatrix} = -25 \cdot \begin{vmatrix} 1 & 3 \\ 3 & 1 \end{vmatrix} + 0 \cdot \begin{vmatrix} 14 & 3 \\ 8 & 1 \end{vmatrix} - 0 \cdot \begin{vmatrix} 14 & 1 \\ 8 & 3 \end{vmatrix} = -25 \cdot [(1)(1) - (3)(3)] + 0 - 0 = 200$$

$$D_y = \begin{vmatrix} 1 & 14 & 3 \\ 2 & 25 & 0 \\ 0 & 8 & 1 \end{vmatrix} = -2 \cdot \begin{vmatrix} 14 & 3 \\ 8 & 1 \end{vmatrix} + 25 \cdot \begin{vmatrix} 1 & 3 \\ 0 & 1 \end{vmatrix} - 0 \cdot \begin{vmatrix} 1 & 14 \\ 0 & 8 \end{vmatrix}$$

$$= -2 \cdot [(14)(1) - (8)(3)] + 25 \cdot [(1)(1) - (0)(3)] - 0 = 20 + 25 = 45$$

$$D_z = \begin{vmatrix} 1 & 1 & 14 \\ 2 & 0 & 25 \\ 0 & 3 & 8 \end{vmatrix} = -2 \cdot \begin{vmatrix} 1 & 14 \\ 3 & 8 \end{vmatrix} + 0 \cdot \begin{vmatrix} 1 & 14 \\ 0 & 8 \end{vmatrix} - 25 \cdot \begin{vmatrix} 1 & 1 \\ 0 & 3 \end{vmatrix}$$

$$= -2 \cdot [(1)(8) - (3)(14)] + 0 - 25 \cdot [(1)(3) - (0)(1)] = 68 - 75 = -7$$

From these determinants, we calculate the solution:

$$x = \frac{D_x}{D} = \frac{200}{16} = 12.5 \quad, \quad y = \frac{D_y}{D} = \frac{45}{16} = 2.8125 \quad \text{and} \quad z = \frac{D_z}{D} = \frac{-7}{16} = -0.4375$$

The solution set of the system is $\{(12.5, 2.8125, -0.4375)\}$.

EXERCISES

Exercises 1–8. Evaluate each determinant.

1. $\begin{vmatrix} 4 & 7 \\ 3 & 7 \end{vmatrix}$

2. $\begin{vmatrix} 5 & 6 \\ 1 & 10 \end{vmatrix}$

3. $\begin{vmatrix} 2 & -3 \\ 6 & 11 \end{vmatrix}$

4. $\begin{vmatrix} 9 & 2 \\ 3 & -4 \end{vmatrix}$

5. $\begin{vmatrix} 3 & -7 & 0 \\ 2 & 6 & -1 \\ -5 & 1 & 0 \end{vmatrix}$

6. $\begin{vmatrix} 2 & 5 & 13 \\ -3 & -7 & 11 \\ 0 & 0 & 4 \end{vmatrix}$

7. $\begin{vmatrix} 10 & 1 & 2 \\ 9 & 0 & 3 \\ 4 & -1 & -5 \end{vmatrix}$

8. $\begin{vmatrix} 3 & 8 & 6 \\ 7 & 6 & 7 \\ 1 & 8 & 2 \end{vmatrix}$

*Exercises 9-14. For each system, set up but **do not** evaluate the determinants $D, D_x, D_y,$ and D_z (if applicable).*

9. $\begin{cases} -x + 5y = 36 \\ 14x - 9y = 18 \end{cases}$

10. $\begin{cases} 4x + 17y = 100 \\ 15x - y = 76 \end{cases}$

11. $\begin{cases} -81x - 37y = 6 \\ 18x - 11y = 36 \end{cases}$

12. $\begin{cases} -4x + 3y - 9z = 21 \\ 2x - y + 7z = 58 \\ 4x + 10y - 3z = 25 \end{cases}$

13. $\begin{cases} 50x - 61y + 2y = 12 \\ -7x + 89y + 93z = -41 \\ 35x - 32y - 85z = 21 \end{cases}$

14. $\begin{cases} 80x + 88y + 30z = 67 \\ -3x + 93y - 29z = 38 \\ 73x - 95y = 40 \end{cases}$

Exercises 15–18. Use Cramer's Rule, if possible, to solve each system of linear equations. Compare your results to those obtained for Exercises 1–4 of Section 4.4 in this Study Guide. See Example 1.

15. $\begin{cases} x + y = -10 \\ x - 6y = 18 \end{cases}$

16. $\begin{cases} x - 3y = 9 \\ x + y = 18 \end{cases}$

17. $\begin{cases} x - 5y = 15 \\ 2x + 5y = 12 \end{cases}$

18. $\begin{cases} x + 4y = -10 \\ 3x - y = -4 \end{cases}$

Exercises 19–22. Use Cramer's Rule, if possible, to solve each system of linear equations. Compare your results to those obtained for Exercises 10–13 of Section 4.4 in this Study Guide. See Example 2.

19. $\begin{cases} x + y + z = 0 \\ 2x + y = 10 \\ z = 2 \end{cases}$

20. $\begin{cases} 2x + z = -8 \\ x + y = 7 \\ y + 3z = -10 \end{cases}$

21. $\begin{cases} x - y - z = 9 \\ x + y + z = 32 \\ 2x - z = 0 \end{cases}$

22. $\begin{cases} -x + y + z = -5 \\ -x + 2z = 4 \\ x + 2y + 3z = 12 \end{cases}$

C *Exercises 23–24. Solve each system of equations using Cramer's Rule. Use the matrix capabilities of a grapher to check each determinant. Use the grapher to check your solution graphically.*

23. $\begin{cases} y = 6.5x - 5.6 \\ y = 2.1x + 0.9 \end{cases}$

24. $\begin{cases} 5.2x + 7.9y = 2.3 \\ 5.5x - 9.1y = 7.3 \end{cases}$

Chapter 4   Hints and Warnings

- To **solve a system of linear equations by the substitution method:**
  - Step 1: Solve one equation for one of the variables.
  - Step 2: Substitute the expression for the variable into the other equation.
  - Step 3: Solve the equation from Step 2 to find the value of one variable.
  - Step 4: Substitute the value from Step 3 in either original equation to find the value of the other variable.
  - Step 5: Check the solution in both equations.

  Drawing a graph of the system of equations can also be helpful when you are verifying the solution.

- To **solve a system of linear equations by the addition method:**
  - Step 1: Rewrite each equation in the standard form $Ax + By = C$.
  - Step 2: Multiply one or both equations by a nonzero number so that the coefficients of a variable are opposites. *Check the system carefully to see if only one equation requires multiplication by a nonzero number. For instance, if the two equations are $3x - 2y = 0$ and $2x + y = 10$, it is easier to multiply the second equation by 2 so that the y-coefficients are opposites than to multiply the first equation by 2 and the second equation by –2 so that the x-coefficients are opposites.*
  - Step 3: Add the equations.
  - Step 4: Find the value of one variable by solving the resulting equation.
  - Step 5: Substitute the value from Step 4 into either original equation to find the value of the other variable.
  - Step 6:   Check the solution in both equations.

- To **solve a system of three linear equations by the elimination method:**
  - Step 1: Write each equation in standard form $Ax + By + Cz = D$.
  - Step 2: Choose a pair of equations and use the equations to eliminate a variable.

Step 3: Choose any other pair of equations and eliminate the same variable.
Step 4: Solve the system of two equations in two variables from Steps 1 and 2.
Step 5: Solve for the third variable by substituting the values of the variables from Step 4 into any of the original equations.

Watch for any equations in which two of the three variables are already eliminated. In that case, find the value of the remaining variable and substitute that value into the other two equations. Now you need only solve a system of two equations to find the values of the other two variables.

- To **perform elementary row operations on a matrix,** you may:
  1. Interchange any two rows.
  2. Multiply (or divide) all the elements of one equation by the same nonzero number.
  3. Multiply (or divide) the elements of one row by the same nonzero number and add it to its corresponding elements in another row.

CHAPTER 4   Practice Test

Evaluate each determinant.

1. $\begin{vmatrix} 3 & 1 \\ 4 & -4 \end{vmatrix}$

2. $\begin{vmatrix} 6 & 6 & 9 \\ -3 & 4 & 0 \\ 5 & 4 & -7 \end{vmatrix}$

Solve each system of equations graphically and then solve by the addition method or the substitution method.

3. $\begin{cases} 2x + 3y = -8 \\ 4x + 2y = -8 \end{cases}$

4. $\begin{cases} -5x + 3y = 6 \\ 10x - 6y = 15 \end{cases}$

Solve each system.

5. $\begin{cases} 4x - 3y = 12 \\ 5x + 10y = 15 \end{cases}$

6. $\begin{cases} 7x - 7y = 15 \\ -14x + 14y = -30 \end{cases}$

7. $\begin{cases} x + 2z = 0 \\ \dfrac{1}{2}x + 3y + z = 0 \\ 3z = -6 \end{cases}$

8. $\begin{cases} x + y + z = 7 \\ 4y - z = 12 \\ 3z = 24 \end{cases}$

9. $\begin{cases} 5x + 2y = 12 \\ 10x + \dfrac{1}{4}y = -6 \end{cases}$

Solve each system using Cramer's Rule.

10. $\begin{cases} x + y = -1 \\ 2x - y = 16 \end{cases}$

11. $\begin{cases} 3x - y = 10 \\ 5x + 2y = 2 \end{cases}$

12. $\begin{cases} x + y + z = -4 \\ 5y = 25 \\ x - y - z = -4 \end{cases}$

13. $\begin{cases} x - 2y + z = -5 \\ x + y = 13 \\ 3x - 2y = 9 \end{cases}$

Solve each system using matrices.

14. $\begin{cases} x + 5y = 7 \\ 2x + 10y = 3 \end{cases}$

15. $\begin{cases} x - y = 1 \\ 3x - 2y = -5 \end{cases}$

16. $\begin{cases} x + y + z = 1 \\ 2x + y = -3 \\ 2x + 3y + 2z = -5 \end{cases}$

17. $\begin{cases} 2x + y - z = -4 \\ 2x + 3z = 19 \\ 3x + 2y = 7 \end{cases}$

18. A sporting goods store bought $3000 worth of equipment for personalizing sports jerseys. Supplies for personalization cost $2.50 per jersey, and the store plans to charge $8.50 per jersey for personalization. Find the number of jerseys that must be personalized to break even.

19. A motel in Rochester, New York, charges $59 including tax per night for a room on a weekend. During the week, rooms cost $79 including tax per night. If the total receipts for one week were $31,955 and a total of 445 rooms were rented, how many rooms were rented during the week and how many were rented on the weekend?

20. A bulk food store blends its own trail mix. The Deluxe Trail Mix sells for $6.98 per pound. How many pounds of raisins (costing $2.99 per pound) must be added to 10 pounds of Deluxe Trail Mix to make Budget Trail Mix costing $4.50 per pound?

C H A P T E R 5

Polynomial Functions

5.1   Exponents and Scientific Notation

EXAMPLE 1   *Objective 1: Use the product rule for exponents.*

Use the product rule to simplify.
  a. $(-2x^4z^{12})(5xy^2z^3)$ **b.** $(4^2p)(4p^3)(rp^5)$

Solution: a. $(-2x^4z^{12})(5xy^2z^3) = (-2)(5)x^4x^1y^2z^{12}z^3 = -10x^{4+1}y^2z^{12+3} = -10x^5y^2z^{15}$
  b. $(4^2p)(4p^3)(rp^5) = 4^24^1p^1p^3p^5r^1 = 4^{2+1}p^{1+3+5}r = 4^3p^9r = 64p^9r$

EXAMPLE 2   *Objective 4: Define a raised to the negative nth power.*

Simplify each expression. Write answers using positive exponents.

  a. $y^{-6}y^3$ **b.** $\dfrac{2a^2}{2^3a^5b^{-3}}$ **c.** $\dfrac{s^{-3}t^5}{rs^2t^2}$ **d.** $\dfrac{(7x^{-7})(8^{-2}x^3y^{-4})}{7^{-1}x^{-2}y^{-3}}$

Solution: a. $y^{-6}y^3 = y^{-6+3} = y^{-3} = \dfrac{1}{y^3}$

  b. $\dfrac{2a^2}{2^3a^5b^{-3}} = 2^{1-3}a^{2-5}b^{-(-3)} = 2^{-2}a^{-3}b^3 = \dfrac{b^3}{4a^3}$

  c. $\dfrac{s^{-3}t^5}{rs^2t^2} = r^{-1}s^{-3-2}t^{5-2} = r^{-1}s^{-5}t^3 = \dfrac{t^3}{rs^5}$

  d. $\dfrac{(7x^{-7})(8^{-2}x^3y^{-4})}{7^{-1}x^{-2}y^{-3}} = 7^{1-(-1)}8^{-2}x^{-7+3-(-2)}y^{-4-(-3)} = 7^28^{-2}x^{-2}y^{-1} = \dfrac{49}{64x^2y}$

EXAMPLE 3   *Objective 5: Write numbers in scientific notation.*

Write each number in scientific notation.
  a.   4,290,000   **b.**   0.0006783

Solution: a.   *Step 1*   Move the decimal point until the number is between 1 and 10.
                              4.290000
             Step 2   The decimal point is moved to the left 6 places, so the count is positive 6.
             Step 3 $4{,}290{,}000 = 4.29 \times 10^6$.
  b.   *Step 1*   Move the decimal point until the number is between 1 and 10.
                              00006.783
             Step 2   The decimal point is moved to the right 4 places, so the count is negative 4.
             Step 3 $0.0006783 = 6.783 \times 10^{-4}$.

EXERCISES

Exercises 1–6. Use the product rule to simplify each expression. See Example 1.

1. $5^2 \cdot 5$ **2.** $(-2)^3(-2)^2$ **3.** $h^2 \cdot h^6$ **4.** $y^4 \cdot y^8$ **5.** $x^3 \cdot x$ **6.** $p \cdot p^5$

Exercises 7–10. Evaluate the following.

7. 2^0 **8.** $(-3)^0$ **9.** q^0 **10.** $3 - 7y^0$

Exercises 11–18. Simplify each expression. Write answers using positive exponents. See Example 2.

11. $(-3)^{-3}$ **12.** 5^{-4} **13.** $4a^{-2}a^{-3}$ **14.** $6s^{-6}(5s^6)$

15. $a^5b^{-5}(a^{-7}b^{-3})$ **16.** $(10x^{-3}y^2)(2^{-1}x^9y^{-4})$ **17.** $\dfrac{(4w^{-1})w^5}{w^8}$ **18.** $\dfrac{24q^{-2}r^6}{3q^{-3}r^9}$

Exercises 19–26. Write each number in scientific notation. See Example 3.

19. 5,744,000,000   **20.** 33,200   **21.** 0.00461   **22.** 0.00000001218

23. 0.002   **24.** 0.000077521   **25.** 90,000,000   **26.** 280

Exercises 27–34. Write each number in standard decimal notation, without exponents.

27. 6.7×10^3 **28.** 3.992×10^7 **29.** 3.221×10^{-4} **30.** 5.7×10^{-1}

31. 7.21×10^{-5} **32.** 1.947×10^{-3} **33.** 6.074×10^5 **34.** 1.6998×10^2

Exercises 35–38. Write each number in scientific notation.

35. As of May 9, 1996, the public debt of the United States was approximately $5,088,800,000,000. (data from U. S. Department of the Treasury)

36. The interest rate on U.S. savings bonds for May through October 1996 in decimal form was 0.0463. (data from U. S. Department of the Treasury)

37. The total amount of deposits in all U.S. savings institutions in 1925 was \$51,641,000,000. (data from Federal Reserve System)

38. The average number of fatal U.S. airline accidents per departure in 1994 was 0.00000054. (data from National Transportation Safety Board)

Exercises 39–42. Write each number in standard decimal notation, without exponents.

39. The number of U.S. airline departures in 1994 was 7.5×10^6. (data from National Transportation Safety Board)

40. The Federal Reserve Board discount rate was raised to 5.25×10^{-2} in February 1995. (data from Federal Reserve Board)

41. The total spent on advertising in the United States in 1994 was $\$5.5 \times 10^{10}$. (data from Competitive Media Reporting and Publishers Information Bureau)

42. The probability of getting a royal flush in a hand of poker is approximately 1.539×10^{-6}.

C *Exercises 43–45. Use a scientific calculator to perform each operation indicated.*

43. Find the increase in the public debt of United States from $\$3.801 \times 10^{12}$ at the end of 1991 to $\$4.989 \times 10^{12}$ at the end of 1995. (Source: U.S. Department of the Treasury)

44. Divide 6.603×10^8 by 3.58×10^4.

45. Divide 9.29×10^7 by 6.76×10^{48}.

5.2   More Work with Exponents and Scientific Notation

EXAMPLE 1 - *Objective 2: Use exponent rules and definitions to simplify exponential expressions.*

Simplify each expression. Write answers using positive exponents.

a. $(2x^2y^3)^3$ **b.** $\left(\dfrac{w^{-3}}{w^{-2}z^{-1}}\right)^{-1}$ **c.** $\dfrac{(3s)^2(s^5t)^{-1}}{(3t)^3}$

Solution: a. $(2x^2y^3)^3 = 2^3(x^2)^3(y^3)^3 = 8x^{2\cdot3}y^{3\cdot3} = 8x^6y^9$

b. $\left(\dfrac{w^{-3}}{w^{-2}z^{-1}}\right)^{-1} = \dfrac{(w^{-3})^{-1}}{(w^{-2})^{-1}(z^{-1})^{-1}} = \dfrac{w^3}{w^2z} = w^{3-2}z^{-1} = \dfrac{w}{z}$

c. $\dfrac{(3s)^2(s^5t)^{-1}}{(3t)^3} = \dfrac{3^2 s^2(s^5)^{-1}t^{-1}}{3^3 t^3} = 3^{2-3}s^{2-5}t^{-1-3} = 3^{-1}s^{-3}t^{-4} = \dfrac{1}{3s^3t^4}$

EXAMPLE 2   *Objective 3: Compute using scientific notation.*

Perform the indicated operations. Write each result in scientific notation.

a. $(7.6 \times 10^{-8})(9.8 \times 10^{-2})$ **b.** $\dfrac{9.0 \times 10^{-6}}{4.5 \times 10^{-8}}$

Solution: a. Use the rules of exponents to simplify this product and write it in scientific notation.

$$(7.6 \times 10^{-8})(9.8 \times 10^{-2}) = 7.6 \times 9.8 \times 10^{-8} \times 10^{-2}$$
$$= 74.48 \times 10^{-10}$$
$$= (7.448 \times 10^{1}) \times 10^{-10}$$
$$= 7.448 \times 10^{-9}$$

b. Again, apply the rules of exponents.

$$\frac{9.0 \times 10^{-6}}{4.5 \times 10^{-8}} = \left(\frac{9.0}{4.5}\right)\left(\frac{10^{-6}}{10^{-8}}\right) = 2.0 \times 10^{2}$$

EXERCISES

Exercises 1–10. Simplify. Write each answer using positive exponents. See Example 1.

1. $(2^2)^3$ **2.** $(4^{-1})^2$ **3.** $(z^2)^5$ **4.** $(p^{-3})^{-4}$ **5.** $(3q^6)^2$

6. $(2w^{-5})^{-2}$ **7.** $(3xy^7)^3$ **8.** $(a^{-3}b^4c^7)^6$ **9.** $\dfrac{3x^2y^{-4}z^3}{(xyz)^{-3}}$ **10.** $\dfrac{(4x)^{-4}z^6}{x^{-3}y^{-2}z^8}$

Exercises 11–22. Perform the indicated operations. Write each result in scientific notation, rounded to two decimal places. See Example 2.

11. $(3 \times 10^2)(1.7 \times 10^7)$ **12.** $(7 \times 10^3)(3.3 \times 10^8)$ **13.** $(5.9 \times 10^{-5})(7.5 \times 10^6)$

14. $(6.3 \times 10^3)(5.1 \times 10^{-9})$ **15.** $(9 \times 10^{-2})^2$ **16.** $(4 \times 10^8)^3$

17. $\dfrac{8.7 \times 10^6}{4.9 \times 10^7}$ **18.** $\dfrac{2.6 \times 10^7}{9.8 \times 10^4}$ **19.** $\dfrac{1.02 \times 10^{-4}}{3.01 \times 10^{-3}}$

20. $\dfrac{3.15 \times 10^7}{1.43 \times 10^{-2}}$ **21.** $\dfrac{(9 \times 10^{-3})(5.7 \times 10^4)}{3.9 \times 10^7}$ **22.** $\dfrac{(7.8 \times 10^5)(1.5 \times 10^{-6})}{4.3 \times 10^{-5}}$

Exercises 23–26. Simplify. Write each result in scientific notation.

23. $\dfrac{0.0000018}{0.0003}$ **24.** $\dfrac{0.000000081}{27}$ **25.** $\dfrac{0.007 \times 1000}{0.0005}$ **26.** $\dfrac{0.027}{0.003 \times 100}$

Exercises 27–30. Solve.

27. The density D of an object is equivalent to the quotient of its mass M and volume V. Thus $D = \dfrac{M}{V}$. Express in scientific notation the density of an object whose mass is 88,000 pounds and whose volume is 0.11 cubic feet.

28. Use the formula for density given in Exercise 27. Express in scientific notation the density of an object whose mass is 9,500,000 pounds and whose volume is 4000 cubic feet.

29. At the end of 1990, the public debt of the United States was $\$3.36 \times 10^{12}$, and the population of the United States was 2.49×10^{8}. Find the amount of public debt per person in 1990, and express the result in scientific notation. (data from U.S. Department of the Treasury and U.S. Bureau of the Census)

30. At the end of 1994, the public debt of the United States was $\$4.8 \times 10^{12}$, and the population of the United States was 2.60×10^{8}. Find the amount of public debt per person in 1994, and express the result in scientific notation. (data from U.S. Department of the Treasury and U.S. Bureau of the Census)

5.3   Polynomials and Polynomial Functions

EXAMPLE 1   *Objective 3: Define polynomial functions.*

If $P(x) = 2x^3 - 5x^2 + x - 10$, find the following:

  a. $P(2)$ **b.** $P(0)$ **c.** $P(-3)$

Solution: a. Substitute 2 for x in $P(x)$ and simplify.

$$P(x) = 2x^3 - 5x^2 + x - 10$$
$$P(2) = 2(2)^3 - 5(2)^2 + (2) - 10$$
$$= 2(8) - 5(4) + 2 - 10$$
$$= 16 - 20 + 2 - 10$$
$$= -12$$

  b. $P(x) = 2x^3 - 5x^2 + x - 10$
$$P(0) = 2(0)^3 - 5(0)^2 + 0 - 10 = -10$$

  c. $P(x) = 2x^3 - 5x^2 + x - 10$
$$P(-3) = 2(-3)^3 - 5(-3)^2 + (-3) - 10 = -112$$

EXAMPLE 2   *Objectives 4 and 5: Add polynomials and subtract polynomials.*

Perform the indicated operations.

  a. $(7x^3 - 8x^2 + 9x + 5) + (-8x^3 - 6x^2 - 4x + 1)$

  b. $(9y^3 - 4y^2 - y - 3) - (6y^3 - y^2 - 2y + 7)$

Solution: a. To add, remove the parentheses and group like terms.

$$(7x^3 - 8x^2 + 9x + 5) + (-8x^3 - 6x^2 - 4x + 1)$$
$$= 7x^3 - 8x^2 + 9x + 5 - 8x^3 - 6x^2 - 4x + 1$$
$$= 7x^3 - 8x^3 - 8x^2 - 6x^2 + 9x - 4x + 5 + 1 \quad \text{Group like terms.}$$
$$= -x^3 - 14x^2 + 5x + 6 \quad \text{Combine like terms.}$$

b. To subtract, change the sign of each term of the second polynomial and add the result to the first polynomial.
$$(9y^3 - 4y^2 - y - 3) - (6y^3 - y^2 - 2y + 7)$$
$$= 9y^3 - 4y^2 - y - 3 - 6y^3 + y^2 + 2y - 7 \quad \text{Change signs and add.}$$
$$= 9y^3 - 6y^3 - 4y^2 + y^2 - y + 2y - 3 - 7 \quad \text{Group like terms.}$$
$$= 3y^3 - 3y^2 - y - 10 \quad \text{Combine like terms.}$$

EXERCISES

Exercises 1–4. List each term of the following polynomials and state the degree of each term. Find the degree of each polynomial.

1. $7x^6 - 2x^9$ **2.** $3x^4 - x^3 + 10x^2 + 12$ **3.** $1 - y^2 + y^5$ **4.** $2x^2 - 1$

Exercises 5–10. If $P(x) = 3x^2 - 5x + 20$ and $Q(x) = 5x^3 - 2$, find the following. See Example 1.

5. $P(0)$ **6.** $P(3)$ **7.** $P(-1)$ **8.** $Q(-2)$ **9.** $Q(3)$ **10.** $Q(0)$

Exercises 11–14. Solve.

11. The polynomial function $P(x) = -0.22x^2 + 45.7x - 2031.3$ models the relationship between the daily high temperature x in degrees Fahrenheit and the attendance $P(x)$ at a town swimming pool. Find $P(82)$, the number of people who go to the pool when it is 82° Fahrenheit.

12. The tallest structure in the United States is a TV tower in Blanchard, North Dakota. Neglecting air resistance, the polynomial function $P(t) = -16t^2 + 2063$ models the height $P(t)$ of an object dropped from the top of this tower, where t is measured in seconds. Find the height of the object when $t = 2$ and $t = 4$ seconds. (data from U.S. Geological Survey)

13. A car rental company charges a rental fee and a mileage fee for each car rental. The total revenues of this rental company are given by the polynomial function $R(x) = 50 + 0.3x$, where x is the total number of miles that rental cars are driven. Find the total revenue when rental cars are driven 60,000 miles.

14. An object is thrown upward with an initial velocity of 40 feet per second from the TV tower mentioned in Exercise 12. The polynomial function that models its height is $P(t) = -16t^2 + 40t + 2063$. Find the height of the object when $t = 2$, $t = 4$, and $t = 6$ seconds.

Exercises 15–24. Simplify by combining like terms. See Example 2.

15. $(5x^2 - 3) + (2x^2 + 7)$ **16.** $(6 - y^3) + (3y^3 + 7y - 4)$ **17.** $2(x + 7) + (x^2 - 5x + 3)$

18. $(x^3 - 5x) + 5(x^3 + x)$ **19.** Add $(4w^3 - 2w^2 + 10w - 1)$ and $(-2w^3 + 6w^2 - 5w - 7)$

20. $(3x + 14) - (x - 7)$ **21.** $(2x^2 + x - 6) - (x^2 - 5x + 17)$ **22.** $(5x^3 - 12x + 9) - 2(x^3 - 2)$

23. $(19 - 3y - 7y^2) - (y^3 - y^2 + 11y - 1)$ **24.** Subtract $(9w - 7)$ from $(w^2 + 12w - 5)$

C *Exercises 25–26. Perform the indicated operations. Then check the result graphically with a grapher.*

25. $(4.7x^2 - 9.4x + 2.4) + (4.6x^2 + 0.9x - 5.3)$ **26.** $(8x^3 - 0.3x + 7.9) - (-1.3x^2 + 8.4x - 1.8)$

5.4   Multiplying Polynomials

EXAMPLE 1   *Objective 2: Multiply binomials.*

Use the FOIL order to multiply $(5x + 7)(4x - 3)$.

Solution: Remember that FOIL stands for **First**—**Outer**—**Inner**—**Last**.

$$(5x + 7)(4x - 3) = \overset{First}{5x \cdot 4x} + \overset{Outer}{5x(-3)} + \overset{Inner}{7 \cdot 4x} + \overset{Last}{7(-3)}$$

$$= 20x^2 - 15x + 28x - 21$$

$$= 20x^2 + 13x - 21 \qquad \text{Combine like terms.}$$

EXAMPLE 2   *Objective 3: Square a binomial.*

Find the product: $(2q - 13)^2$.

Solution: $(2q - 13)^2 = (2q)^2 - 2(2q)(13) + 13^2$

$$= 2^2 q^2 - 52q + 169$$

$$= 4q^2 - 52q + 169$$

EXAMPLE 3   *Objective 4: Multiply the sum and difference of two terms.*

Find the product: $(7w + 11)(7w - 11)$.

Solution: $(7w + 11)(7w - 11) = (7w)^2 - 11^2 = 49w^2 - 121$

EXERCISES

Exercises 1–4. Multiply.

1. $(7x^6)(-2x^9)$ **2.** $(3x^4y)(10x^2y^2)$ **3.** $(12st)(-2st)$ **4.** $(2x^2)(-xy)$

Exercises 5–8. Multiply the polynomials.

5. $5t(t-3)$ **6.** $12z(1-z)$ **7.** $-2x(ax+2x^2-4)$ **8.** $b(3-2x)$

Exercises 9–14. Multiply the binomials. See Example 1.

9. $(x+2)(x-6)$ **10.** $(d-7)(d+1)$ **11.** $(3x+1)(4x+6)$

12. $\left(\dfrac{1}{2}m+2\right)\left(\dfrac{1}{2}m-10\right)$ **13.** $(2y+4x)(5y-x)$ **14.** $(6r-4)(8r-1)$

Exercises 15–22. Multiply using special product methods. See Examples 2 and 3.

15. $(t-3)^2$ **16.** $(w+6)^2$ **17.** $(4v-2)(4v+2)$ **18.** $(2q+9)(2q-9)$

19. $(2z+5)^2$ **20.** $(4y-10)^2$ **21.** $(3a+8b)(3a-8b)$ **22.** $(2x-y)(2x+y)$

Exercises 23–26. If $R(x)=3x-1$, $P(x)=x^2+2x+1$, and $Q(x)=-4x$, find the following.

23. $Q(x)\cdot R(x)$ **24.** $Q(x)\cdot P(x)$ **25.** $R(x)\cdot P(x)$ **26.** $[R(x)]^2$

27. Find the volume of the cylinder. Do not approximate π.

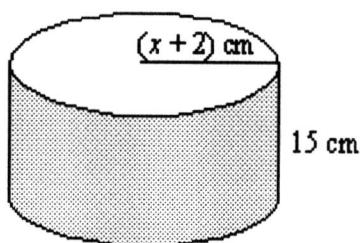

28. Find the area of the circle. Do not approximate π.

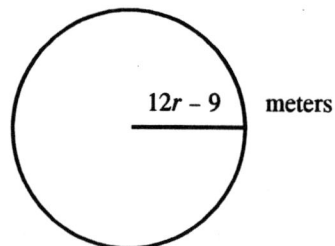

C *Exercises 29–31. Multiply. Then use a grapher to check the results graphically.*

29. $(6x-1)^2$ **30.** $(2x+5)(2x-5)$ **31.** $(3x-1)(x^2+5x-4)$

5.5   The Greatest Common Factor and Factoring by Grouping

EXAMPLE 1   *Objective 2: Factor out the GCF of a polynomial's terms.*

Factor.
  a. $3x^3-6ax^2+27ax$ **b.** $3x(a-b)+2y(a-b)$

Solution: a.   The GCF of the three terms $3x^3$, $-6ax^2$, and $27ax$ is $3x$.
$$3x^3-6ax^2+27ax = 3x(x^2)-3x(2ax)+3x(9a)$$
$$= 3x(x^2-2ax+9a)$$
  b.   The GCF is the binomial factor $(a-b)$. Thus,
$$3x(a-b)+2y(a-b)=(a-b)(3x+2y)$$

EXAMPLE 2   *Objective 3: Factor polynomials by grouping.*

Factor: $2w^3+w^2z-6w-3z$.

Solution: The GCF of all four terms is 1. Try grouping the first two terms together and the last two terms together.
$$2w^3+w^2z-6w-3z = (2w^3+w^2z)+(-6w-3z)$$

Factor w^2 from the first group and -3 from the second group.
$$= w^2(2w+z)-3(2w+3z) \qquad \text{Watch the signs!}$$
$$= (2w+z)(w^2-3) \qquad \text{Factor out a GCF of } (2w+3z).$$

EXERCISES

Exercises 1–4. Find the GCF of each list of monomials.

1. $8x^6, -2x^9, 4x^3$ **2.** $3x^4, 10x^2, 2x$ **3.** $12s^2t, -2st, 8st^3$ **4.** $5q^2r^5, 10q^3r^4, 5q^5r^3$

Exercises 5–12. Factor out the GCF in each polynomial. See Example 1.

5. $5t^2-3t$ **6.** $12z-12z^2$ **7.** $-2ax^2-4ax^3+8x$ **8.** $12b-8bx$

9. $5b(x+2)+3(x+2)$ **10.** $4x(d-7)-y(d-7)$

11. $4x(3x+1)+5(3x+1)$　　　　　　**12.** $12w^2y^3 - 3w^2y + 9w^2y^2$

Exercises 13–20. Factor each polynomial by grouping. See Example 2.

13. $xy + 2x + y + 2$　　　**14.** $ab + 3a - 3b - 9$　　　**15.** $x^2 - 2x + xy - 2y$　　　**16.** $a^2 + ab - 12a - 12b$

17. $21xy - 35x + 6y - 10$　　　　　　**18.** $6xy - 2x - 3y + 1$

19. $2x^2 + x - 2xy - y$　　　　　　**20.** $a^2 - 2a + 2ab - 4b$

21. The area of the given trapezoid is $\dfrac{1}{2}ah + \dfrac{1}{2}bh$. Factor this expression.

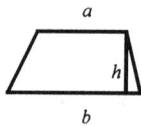

(figure: trapezoid with top side a, height h, bottom side b)

22. The volume of a cardboard box is $2x^2y - x^2 + 4xy - 2x$, where the height is x, the length is $(x + 2)$ and the width is $(2y - 1)$. Factor this expression.

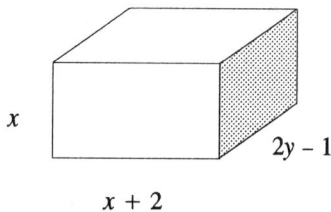

(figure: cardboard box with dimensions x, $2y - 1$, and $x + 2$)

C *Exercises 23–24. Factor. Then use a grapher to check the results graphically.*

23. $3x^3 - 6x^2 + 5x - 10$　　　　　　**24.** $x^4 + 3x^3 - 3x - 9$

5.6   Factoring Trinomials

EXAMPLE 1   *Objective 1: Factor trinomials of the form* $x^2 + bx + c$.

Factor: $3w^3 - 3w^2 - 18w$.

Solution: First, factor out a GCF of $3w$.
$$3w^3 - 3w^2 - 18w = 3w(w^2 - w - 6)$$

Next, factor $w^2 - w - 6$ by finding two numbers whose product is –6 and whose sum is –1. The numbers are –3 and 2 (because $(-3)(2) = -6$ and $-3 + 2 = -1$).
$$3w^3 - 3w^2 - 18w = 3w(w^2 - w - 6)$$
$$= 3w(w - 3)(w + 2)$$

EXAMPLE 2   *Objective 2: Factor trinomials of the form* $ax^2 + bx + c$.

Factor: $8x^2 y + 8xy - 6y$.

Solution: First, factor out the GCF of the terms of this trinomial, $2y$.
$$8x^2 y + 8xy - 6y = 2y(4x^2 + 4x - 3)$$

Now try to factor the trinomial $4x^2 + 4x - 3$. Factors of $4x^2$ are $4x^2 = 2x \cdot 2x$ and $4x^2 = 4x \cdot 1x$. Try $2x$ and $2x$.
$$2y(4x^2 + 4x - 3) = 2y(2x + \quad)(2x + \quad)$$

The constant term –3 is negative and the coefficient of the middle term 4 is positive, so factor –3 into positive and negative factors. The factors of –3 are $-3 = (-3)(1)$ and $-3 = (3)(-1)$. Try 3 and –1.

$$2y(2x+3)(2x-1)$$

$6x$

$\frac{-2x}{4x}$ correct middle term

If this combination had not worked, we would try –3 and 1. The final result is
$$2y(4x^2 + 4x - 3) = 2y(2x + 3)(2x - 1)$$

EXERCISES

Exercises 1–6. Factor each trinomial. See Example 1.

1. $x^2 - 3x + 2$

2. $x^2 - 4x - 5$

3. $x^2 + 8x + 15$

4. $x^2 + 4x - 21$

5. $2x^2 + 36x + 160$

6. $3x^2 - 3x - 126$

Exercises 7–12. Factor each trinomial. See Example 2.

7. $2x^2 + 3x + 1$

8. $2x^2 - 17x - 9$

9. $6x^2 - 11x - 10$

10. $10x^2 - x - 3$

11. $8x^2 - 42x - 36$

12. $27x^2 y + 9xy - 6y$

Exercises 13–20. Factor each polynomial completely.

13. $x^2 - 16x + 63$

14. $5st^2 - 80st + 315s$

15. $2x^2 + 15x - 27$

16. $6w^3 + 45w^2 - 81w$

17. $2(3x - 1)^2 + 9(3x - 1) - 110$

18. $y^4 + 13y^2 + 30$

19. $q^5 - 5q^3 + 4q$

20. $6x^5 + 6x^3 - 36x$

21. The volume of a cardboard box in terms of its width x is $V(x) = 9x^3 - 18x^2 + 8x$.

(a) Factor this expression for $V(x)$.

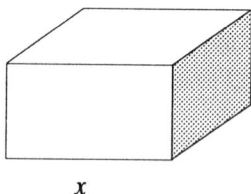

(b) If the width of the box is 7 centimeters, use your results from part (a) to find the length and height of the box. What is the volume of this box?

x

22. Suppose you are a math instructor and you have just given a quiz on factoring. One of your students gives $(x - 3)(2x + 8)$ as his answer to the question "Factor $2x^2 + 2x - 24$ completely." Is this answer right, wrong, or partially wrong? Explain.

C *Exercises 23–24. Factor completely. Then use a grapher to check the results graphically.*

23. $5x^5 + 10x^4 + x^3$

24. $6x^3 + 25x^2 + 24x$

5.7   Factoring by Special Products and Factoring Strategies

EXAMPLE 1   *Objective 1: Factor a perfect square trinomial.*

Factor: $9w^2 - 12w + 4$.

Solution: Notice that the first term is a perfect square: $(3w)^2 = 9w^2$, the last term is a perfect square: $2^2 = 4$, and $12w = 2 \cdot 2 \cdot 3w$. Thus,
$$9w^2 - 12w + 4 = (3w)^2 - 2(3w)(2) + 2^2 = (3w - 2)^2.$$

EXAMPLE 2   *Objective 2: Factor the difference of two squares.*

Factor: $75y^2 - 363$.

Solution: First, factor out the GCF of the terms of this binomial, 3.
$$75y^2 - 363 = 3(25y^2 - 121)$$

Now, the binomial $25y^2 - 121$ is the difference of two squares.
$$75y^2 - 363 = 3(25y^2 - 121) = 3[(5y)^2 - 11^2] = 3(5y - 11)(5y + 11).$$

EXAMPLE 3   *Objective 3: Factor the sum or difference of two cubes.*

Factor: $27q^3 - \dfrac{r^3}{8}$.

Solution: This is a difference of cubes because $27q^3 - \dfrac{r^3}{8} = (3q)^3 - \left(\dfrac{r}{2}\right)^3$.

$$(3q)^3 - \left(\dfrac{r}{2}\right)^3 = \left(3q - \dfrac{r}{2}\right)\left((3q)^2 + (3q)\dfrac{r}{2} + \left(\dfrac{r}{2}\right)^2\right)$$

$$= \left(3q - \dfrac{r}{2}\right)\left(9q^2 + \dfrac{3qr}{2} + \dfrac{r^2}{4}\right).$$

EXERCISES

Exercises 1–6. Factor the following. See Example 1.

1. $x^2 - 8x + 16$ **2.** $9x^2 + 48x + 64$ **3.** $4x^2 - 20x + 25$

4. $2z^2 - 40z + 200$ **5.** $5y^2 - 30y + 45$ **6.** $12x^2 + 84x + 147$

Exercises 7–10. Factor the following. See Example 2.

7. $16w^2 - 9z^2$ **8.** $50x^2z - 18y^2z$ **9.** $36r^2 - 169q^2$ **10.** $3x^2 - 12y^2z^2$

Exercises 11–14. Factor the following. See Example 3.

11. $w^3 - 1000$ **12.** $x^3 + 512$ **13.** $8y^3 + 27x^3$ **14.** $64m^3r^2 - 125n^3r^2$

Exercises 15–20. Factor each polynomial completely.

15. $z^6 - 64$ **16.** $\dfrac{1}{25} - 9w^2$ **17.** $16y^2 - 6xy + \dfrac{9x^2}{16}$

18. $72w^3 + 9$ **19.** $63x^2 + 21x + \dfrac{7}{4}$ **20.** $3a^2b - 108b^3$

C *Exercises 21–22. Factor completely. Then use a grapher to check the results graphically.*

21. $75x^2 - 90x + 27$ **22.** $72x - 98x^3$

5.8   Solving Equations by Factoring and Problem Solving

EXAMPLE 1   *Objective 1: Solve polynomial equations by factoring.*

Solve: $5x^3 + 10x^2 + 7x = 13x - 3x^2$.

Solution: First, write the equation in standard form; then factor.

$$5x^3 + 10x^2 + 7x = 13x - 3x^2$$

$5x^3 + 10x^2 + 3x^2 + 7x - 13x = 0$ Subtract $13x$ from both sides and add $3x^2$ to both sides.

$5x^3 + 13x^2 - 6x = 0$ Combine like terms.

$x(5x^2 + 13x - 6) = 0$ Factor out the GCF x.

$x(5x - 2)(x + 3) = 0$ Factor.

$x = 0$ $5x - 2 = 0$ $x + 3 = 0$

or $5x = 2$ or $x = -3$

$x = \dfrac{2}{5}$

The solution set is $\left\{-3, 0, \dfrac{2}{5}\right\}$. Check all three solutions in the original equation.

EXERCISES

Exercises 1–10. Solve each equation. See Example 1.

1. $(d - 7)(d + 1) = 0$ **2.** $(3x + 1)(4x + 6) = 0$ **3.** $x^2 - 4x - 5 = 0$

4. $x^2 - 8x + 16 = 0$

5. $x^2 + 4x - 18 = 3$

6. $3x^2 - 3x - 130 = -4$

7. $2x(4x - 21) + 10 = 46$

8. $9x^2 + 36x + 80 = 4(4 - 3x)$

9. $2(3x - 1)^2 + 9(3x - 1) - 110 = 0$

10. $12x^2 + 91x + 147 = 7x$

11. The product of two consecutive even numbers is equal to 288. Find the two numbers.

12. The product of two consecutive odd numbers is equal to 63. Find the two numbers.

13. Suppose you live next to a rectangular park. How much distance do you save by walking directly from point A to point C rather than walking from A to B to C?

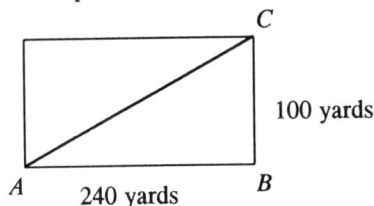

14. Suppose an object is dropped from a height of 784 feet. The height of the falling object after t seconds is given by the polynomial function $h(t) = -16t^2 + 784$. How long will the object be falling through the air?

C *Exercises 15–17. Solve each quadratic equation by graphing a related function with a grapher and approximating the x-intercepts with the Zoom and Trace features.*

15. $2x^2 - 7x + 2 = 0$

16. $0.6x^2 + 2x - 3 = 0$

17. The Washington Monument in Washington, D.C., is 555 feet tall. Suppose a visitor drops a penny from the top of the monument. The height of the falling penny after t seconds is given by the polynomial function $h(t) = -16t^2 + 555$. How long will the penny be falling through the air? (data from National Park Service)

5.9   An Introduction to Graphing Polynomial Functions

EXAMPLE 1   *Objective 3: Find the vertex of a parabola by using the vertex formula.*

Graph: $f(x) = -2x^2 + 13x - 15$.

Solution: To find the vertex, use the vertex formula. For the function $y = -2x^2 + 13x - 15$, $a = -2$ and $b = 13$. Thus,

$$x = \frac{-b}{2a} = \frac{-13}{2(-2)} = \frac{13}{4}$$

Next, find $f\left(\dfrac{13}{4}\right)$.

$$f\left(\frac{13}{4}\right) = -2\left(\frac{13}{4}\right)^2 + 13\left(\frac{13}{4}\right) - 15$$

$$= -\frac{169}{8} + \frac{169}{4} - 15$$

$$= \frac{49}{8}$$

The vertex is $\left(\dfrac{13}{4}, \dfrac{49}{8}\right)$. Also, this parabola opens downward, since $a = -2$, which is less than 0.

Notice that because the vertex lies above the x-axis and opens downward, the parabola has x-intercepts. To find the x-intercepts, let $f(x) = 0$.

$$f(x) = -2x^2 + 13x - 15$$

$0 = -2x^2 + 13x - 15$	Let $f(x) = 0$.
$0 = 2x^2 - 13x + 15$	Divide both sides by -1.
$0 = (2x - 3)(x - 5)$	Factor.

$$2x - 3 = 0 \qquad\qquad x - 5 = 0 \qquad \text{Set each factor equal to 0.}$$

$$2x = 3 \qquad \text{or} \qquad x = 5 \qquad \text{Solve.}$$

$$x = \frac{3}{2}$$

The x-intercepts are $\dfrac{3}{2}$ and 5.

To find the y-intercept, let $x = 0$.

$$f(x) = -2x^2 + 13x - 15$$

$$f(0) = -2(0)^2 + 13(0) - 15$$

$$= -15$$

The y-intercept is -15. Now plot theses points and connect them with a smooth curve.

$f(x) = -2x^2 + 13x - 15$

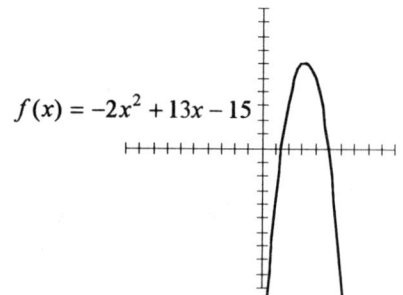

EXERCISES

Exercises 1–4. Find the vertex of the graph of each function.

1. $f(x) = x^2 + 8x + 15$

2. $f(x) = 2x^2 + 36x + 160$

3. $f(x) = 3x^2 - 3x - 126$

4. $f(x) = 8x^2 - 42x - 36$

Exercises 5–6. Find the intercepts of the graph of each function.

5. $f(x) = x^2 + 4x - 21$

6. $f(x) = 12x^2 + 84x + 147$

Exercises 7–12. Graph each quadratic function. See Example 1.

7. $f(x) = -3x^2 + 6x$

8. $f(x) = \dfrac{1}{2}x^2 - 4x$

9. $f(x) = x^2 - 3x + 2$

10. $f(x) = x^2 - 4x - 5$

11. $f(x) = -2x^2 + 3x + 1$

12. $f(x) = x^2 + 3x - 8$

C *Exercises 13–14. Use a grapher to approximate the x-intercepts to the nearest tenths.*

13. $f(x) = -3x^2 - 4x + 3$

14. $f(x) = 0.3x^2 - 4x + 7$

Chapter 5   Hints and Warnings

- To **add polynomials,** combine like terms.
- To **subtract polynomials,** change the signs of the terms of the polynomial being subtracted, then add.
- To **multiply two polynomials,** use the distributive property and multiply each term of one polynomial by each term of the other polynomial, then combine like terms.
- To **multiply two binomials,** the FOIL method may be used. *Remember that FOIL stands for "First terms—Outer terms—Inner terms—Last terms."*
- To **factor a polynomial by grouping,** group the terms so that each group has a common factor. Factor out these common factors. Then see if the new groups have a common factor.
- To **factor** $ax^2 + bx + c$,
    - Step 1: Write all pairs of factors of ax^2.
    - Step 2: Write all pairs of factors of c.
    - Step 3: Try combinations of these factors until the middle term bx is found.
- To **factor a trinomial,**
    - Step 1: Factor out the GCF.
    - Step 2: If the polynomial is a binomial, check to see if it is the difference of two squares, $a^2 - b^2 = (a+b)(a-b)$, or a sum or difference of two cubes, $a^3 + b^3 = (a+b)(a^2 - ab + b^2)$ or $a^3 - b^3 = (a-b)(a^2 + ab + b^2)$. If the polynomial is a trinomial, see if it is a perfect square trinomial, $a^2 + 2ab + b^2 = (a+b)^2$ or $a^2 - 2ab + b^2 = (a-b)^2$. If not, try factoring by the method listed above for $ax^2 + bx + c$. If the polynomial has 4 or more terms, try factoring by grouping.
    - Step 3: See if any factors can be factored further.
- To **solve polynomial equations by factoring:**
    - Step 1: Write the equation so that one side is equal to 0.
    - Step 2: Factor the polynomial completely.
    - Step 3:   Set each factor equal to 0.
    - Step 4: Solve the resulting equations.
    - Step 5: Check each solution in the original equation.
- To **graph a polynomial function,** find and plot x-intercepts, y-intercepts and a sufficient number of ordered pair solutions. Then connect the plotted points with a smooth curve. *If the polynomial is a quadratic function, remember that the vertex of the parabola is given by $\left(\dfrac{-b}{2a}, f\left(\dfrac{-b}{2a}\right) \right)$. You can also use the symmetry of the parabola to help you connect the plotted ordered pair solutions.*

CHAPTER 5   Practice Test

Simplify. Use positive exponents to write answers.

1. $(3y)^{-3}$ **2.** $-2p^2 q^3 (2pq)^{-2}$ **3.** $\dfrac{3m^2 n^4}{9m^{-3} n^7}$ **4.** $\left(\dfrac{2a^{-2} b^{-1} c^3}{4a^4 c^2} \right)^{-1}$

Write in scientific notation.

5.   56,730,000   **6.**   0.000354

Write without exponents.

7. 8.23×10^{-4}

8. 6.19×10^{8}

Perform the indicated operations.

9. $(3x^2 - 15x + 2) - (x^2 + 7x - 9)$

10. $-6ab(2a^2 + b - 3ab)$

11. $(3w - 5z)(2w + 7z)$

12. $(2x - 5)^2$

13. $(3x - 10)(3x + 10)$

14. $(z - 1)(6z^2 + 4z - 1)$

Factor each polynomial completely.

15. $10x^3y^3 - 5xy^4$

16. $x^2 + 14x + 33$

17. $9y^2 - 6y + 1$

18. $8x^3 - 4x^2 - 4x$

19. $9x^2 - 169$

20. $x^3 - 125$

21. $14x^2 + 63$

22. $x^3 + 7x^2 - 4x - 28$

Solve each equation for the variable.

23. $6(w + 9)(2w - 7) = 0$

24. $(2q + 5)(2q - 5) = 3q^2 - 24$

25. $9x^3 - 18x^2 - x + 4 = 2$

26. $4y^2 - 81 = 63$

Graph.

27. $x^2 - 2x - 8$

28. $3x^3 + 2$

CHAPTER 6

Rational Expressions

6.1   Rational Functions and Simplifying Rational Expressions

EXAMPLE 1   *Objective 2: Find the domain of a rational function.*

Find the domain of each rational function.

a. $f(x) = \dfrac{6x^2 - 5x + 17}{13}$ **b.** $g(x) = \dfrac{7x^3 + 12x}{3x + 14}$ **c.** $h(x) = \dfrac{2x^2 - 5x + 9}{x^2 + 3x + 2}$

Solution: The domain of each function will contain all real numbers except those values that make the denominator 0.

a. No matter what the value of x, the denominator of $\dfrac{6x^2 - 5x + 17}{13}$ is never 0, so the domain of f is $\{x|x \text{ is a real number}\}$.

b. To find the values of x that make the denominator of $g(x)$ equal to 0, we solve the equation
$$3x + 14 = 0$$
$$3x = -14$$
$$x = -\frac{14}{3}$$

The domain of $g(x)$ must exclude $-\dfrac{14}{3}$ because the rational expression is undefined when x is $-\dfrac{14}{3}$. The domain of g is $\left\{x|x \text{ is a real number and } x \neq -\dfrac{14}{3}\right\}$.

c. We find the domain by setting the denominator equal to 0.

$$x^2 + 3x + 2 = 0 \qquad \text{Set the denominator equal to 0.}$$
$$(x + 2)(x + 1) = 0 \qquad \text{Factor.}$$
$$x + 2 = 0 \quad \text{or} \quad x + 1 = 0 \qquad \text{Set each factor equal to 0.}$$
$$x = -2 \quad \text{or} \qquad x = -1 \quad \text{Solve.}$$

The domain of h is $\{x|x \text{ is a real number and } x \neq -2 \text{ and } x \neq -1\}$.

EXAMPLE 2   *Objective 3: Write a rational expression in lowest terms.*

Simplify each rational expression.

a. $\dfrac{3x^2 - 15x}{6x^3 - 3x^2 + 12x}$ **b.** $\dfrac{a^2 - a - 12}{2a^2 - 3a - 20}$

Solution: a. Factor the GCF out of the numerator and the denominator.

$$\frac{3x^2 - 15x}{6x^3 - 3x^2 + 12x} = \frac{3x(x - 5)}{3x(2x^2 - x + 4)} = \frac{x - 5}{2x^2 - x + 4}$$

b. Factor the numerator and the denominator.

$$\frac{a^2 - a - 12}{2a^2 - 3a - 20} = \frac{(a + 3)(a - 4)}{(2a + 5)(a - 4)} = \frac{a + 3}{2a + 5}$$

EXAMPLE 3   *Objective 4: Write a rational expression equivalent to a rational expression with a given denominator.*

Write each rational expression as an equivalent rational expression with the given denominator.

a. $\dfrac{5ab}{6c^2}$, denominator $18abc^2$ **b.** $\dfrac{x-2}{3x+2}$, denominator $6x^2-11x-10$

Solution: a. $\dfrac{5ab}{6c^2} = \dfrac{?}{18abc^2}$

If the denominator $6c^2$ is multiplied by $3ab$, the result is the given denominator $18abc^2$. Use the fundamental principle of rational expressions and multiply the numerator and the denominator of the original rational expression by $3ab$. Then

$$\frac{5ab}{6c^2} = \frac{5ab(3ab)}{6c^2(3ab)} = \frac{15a^2b^2}{18abc^2}$$

b.   The factored form of the given denominator, $6x^2-11x-10$, is $(3x+2)(2x-5)$.

$$\frac{x-2}{3x+2} = \frac{?}{6x^2-11x-10} = \frac{?}{(3x+2)(2x-5)}$$

Use the fundamental principle of rational expressions and multiply the numerator and denominator of the original rational expression by $(2x-5)$.

$$\frac{x-2}{3x+2} = \frac{(x-2)(2x-5)}{(3x+2)(2x-5)} = \frac{2x^2-9x+10}{(3x+2)(2x-5)} = \frac{2x^2-9x+10}{6x^2-11x-10}$$

EXERCISES

Exercises 1–2. Find each function value.

1. $f(x) = \dfrac{x^2+3x-10}{5x+4}$, $f(0), f(4)$ **2.** $g(t) = \dfrac{100-3t}{t^3+2t}$, $g(4), g(-1)$

Exercises 3–6. Find the domain of each function. See Example 1.

3. $\dfrac{5q^4-2q+9}{11}$ **4.** $\dfrac{2t^2-t-5}{t+7}$ **5.** $\dfrac{3x^2+25}{2x+3}$ **6.** $\dfrac{4y^3-y^2+3}{y^2-6y+5}$

Exercises 7–10. Write each rational expression in lowest terms. See Example 2.

7. $\dfrac{4r^2+20r}{4r^3-8r^2+8r}$ **8.** $\dfrac{11x-2}{2-11x}$ **9.** $\dfrac{y^2+4y+3}{2y^2-y-3}$ **10.** $\dfrac{4y^2-49}{2y-7}$

Exercises 11–14. Write each rational expression as an equivalent rational expression with the given denominator. See Example 3.

11. $\dfrac{3}{7y^2}$, $7y^3 + 7y^2$ **12.** $\dfrac{2}{y}$, $3y^2 - 5y$ **13.** $\dfrac{x}{x+3}$, $x^2 - 6x - 27$ **14.** $\dfrac{q-3}{3q+1}$, $9q^2 - 1$

15. The annual revenue from the sale of a new mountain bike model is approximated by the rational function

$$R(x) = \frac{80x^2}{2x^2 - 3x + 4}$$

where x is the number of years since the model was first released and $R(x)$ is the annual revenue in hundreds of thousands of dollars.

(a) What is the annual revenue for the first year?
(b) What is the annual revenue for the second year?
(c) What is the annual revenue for the third year?
(d) What is the annual revenue for the fourth year?

C *Exercises 16–17. Find the domain of each rational function. Then use a grapher to graph each rational function and use the graph to confirm the domain.*

16. $f(x) = \dfrac{3x+1}{2x^2 - 7x + 3}$ **17.** $f(x) = \dfrac{4x^2 + 5}{6x^2 + 33x + 27}$

6.2   Multiplying and Dividing Rational Expressions

EXAMPLE 1   *Objective 1: Multiply rational expressions.*

Multiply: $\dfrac{x^2 + 9x + 14}{-3x + 4} \cdot \dfrac{6x^2 - 8x}{x^2 - 49}$.

Solution:

$$\frac{x^2 + 9x + 14}{-3x + 4} \cdot \frac{6x^2 - 8x}{x^2 - 49} = \frac{(x+2)(x+7)}{-1(3x-4)} \cdot \frac{2x(3x-4)}{(x+7)(x-7)} \quad \text{Factor.}$$

$$= \frac{2x(x+2)(x+7)(3x-4)}{-1(3x-4)(x+7)(x-7)} \quad \text{Multiply.}$$

$$= -\frac{2x(x+2)}{x-7} \quad \text{Divide out common factors.}$$

EXAMPLE 2   *Objective 2: Divide by a rational expression.*

Divide: $\dfrac{2a^2 + 9a + 9}{5a - 30} \div \dfrac{4a^2 - 9}{3a^2 - 15a - 18}$.

Solution: To divide, multiply by the reciprocal of the divisor.

$$\frac{2a^2+9a+9}{5a-30} \div \frac{4a^2-9}{3a^2-15a-18} = \frac{2a^2+9a+9}{5a-30} \cdot \frac{3a^2-15a-18}{4a^2-9}$$

$$= \frac{(a+3)(2a+3)}{5(a-6)} \cdot \frac{3(a-6)(a+1)}{(2a+3)(2a-3)}$$

$$= \frac{3(a+3)(a+1)}{5(2a-3)}$$

EXERCISES

Exercises 1–6. Perform the indicated operation. Write all answers in lowest terms. See Examples 1 and 2.

1. $\dfrac{2y-10}{y^2+3y} \cdot \dfrac{y^2}{5-y}$

2. $\dfrac{y+3}{2} \cdot \dfrac{2y+1}{y^2+6y+9}$

3. $\dfrac{3a+6}{a^2-1} \div \dfrac{a^2+4a+4}{4a-4}$

4. $\dfrac{x+2}{2x} \div \dfrac{4x^2-16}{20}$

5. $\dfrac{4y^2+4y-15}{4y^2+20y+25} \cdot \dfrac{y^2-4y-5}{3-2y}$

6. $\dfrac{x^3+x^2+2x}{x^2+4x-5} \div \dfrac{x}{x^2+7x+10}$

Exercises 7–12. If $f(x)=9-3x$ and $g(x)=x^2+x-12$, find the following.

7. $(f+g)(x)$

8. $(g-f)(x)$

9. $(f \cdot g)(x)$

10. $\left(\dfrac{g}{f}\right)(x)$

11. $(g-f)(2)$

12. $(f \cdot g)(0)$

13. Find the area of the triangle, where the lengths of the base and height are given.

14. Find the area of the rectangle.

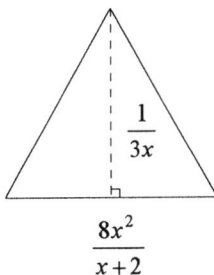

$\dfrac{1}{3x}$

$\dfrac{8x^2}{x+2}$

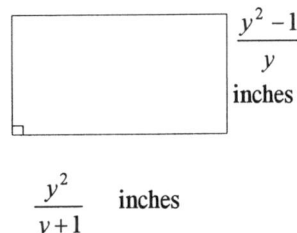

$\dfrac{y^2-1}{y}$ inches

$\dfrac{y^2}{y+1}$ inches

6.3   Adding and Subtracting Rational Expressions

EXAMPLE 1   *Objective 2: Identify the least common denominator of two or more rational expressions.*

Find the LCD of the rational expressions.

$$\frac{3}{a^2b^2}, \frac{a}{5b^3}$$

Solution: Factor each denominator.

$$a^2b^2 = a \cdot a \cdot b \cdot b$$

$$5b^3 = 5 \cdot b \cdot b \cdot b$$

The unique factors are 5, a, and b.
The LCD is $5 \cdot a \cdot a \cdot b \cdot b \cdot b = 5a^2b^3$.

EXAMPLE 2   *Objective 3: Add and subtract rational expressions with unlike denominators.*

Subtract: $\dfrac{x}{x^2 + x - 2} - \dfrac{2x}{x^2 + 2x - 3}$.

Solution: The denominators factor as $x^2 + x - 2 = (x-1)(x+2)$ and $x^2 + 2x - 3 = (x-1)(x+3)$. The LCD is the product of the factors $(x-1)(x+2)(x+3)$.

$$\frac{x}{x^2 + x - 2} - \frac{2x}{x^2 + 2x - 3} = \frac{x}{(x-1)(x+2)} - \frac{2x}{(x-1)(x+3)}$$

$$= \frac{x(x+3)}{(x-1)(x+2)(x+3)} - \frac{2x(x+2)}{(x-1)(x+2)(x+3)}$$

$$= \frac{x(x+3) - 2x(x+2)}{(x-1)(x+2)(x+3)}$$

$$= \frac{x^2 + 3x - 2x^2 - 4x}{(x-1)(x+2)(x+3)}$$

$$= \frac{-x^2 - x}{(x-1)(x+2)(x+3)}$$

$$= \frac{-x(x+1)}{(x-1)(x+2)(x+3)}$$

Because the numerator and denominator have no common factors, this rational expression is in lowest terms.

EXERCISES

Exercises 1–2. Perform the indicated operation. Write each answer in lowest terms.

1. $\dfrac{10}{5y - y^2} - \dfrac{y+5}{5y - y^2}$

2. $\dfrac{5y - 3}{y^2 + 10y + 9} + \dfrac{4 - 4y}{y^2 + 10y + 9}$

Exercises 3–6. Find the LCD of the rational expressions in the list. See Example 1.

3. $\dfrac{3}{x+5}, \dfrac{2x}{5-x}$

4. $\dfrac{1}{xy}, \dfrac{a}{12xy^2z}$

5. $\dfrac{t}{t^2-16}, \dfrac{11}{t+4}$

6. $\dfrac{z}{z^2-7z+6}, \dfrac{z+3}{z^2-6z}$

Exercises 7–12. Perform the indicated operation. Write each answer in lowest terms. See Example 2.

7. $\dfrac{x}{x+3} + \dfrac{4}{x+4}$

8. $\dfrac{5x}{4x^2-9} - \dfrac{7x-3}{2x+3}$

9. $\dfrac{10}{x} - \dfrac{2}{x-1}$

10. $\dfrac{3}{3x^2-5x-2} + \dfrac{x-1}{3x^2+10x+3}$

11. $\dfrac{2x}{x-1} + \dfrac{3}{x+1} - \dfrac{2x}{x^2-1}$

12. $\dfrac{11}{(x-2)^2} - \dfrac{3x}{x^2-4}$

C *Exercises 13–14. Perform the indicated operation. Then use a grapher to confirm your work by graphing the original problem and your simplified answer on the same screen.*

13. $\dfrac{x-1}{6x^2+10x-4} - \dfrac{x+2}{3x^2-4x+1}$

14. $\dfrac{5}{2x+4} - \dfrac{x}{x+2} + \dfrac{2x}{-3x-6}$

6.4   Simplifying Complex Fractions

EXAMPLE 1   *Objective 2: Simplify complex fractions by simplifying the numerator and denominator and then dividing.*

Simplify the complex fraction.

$$\dfrac{2a - \dfrac{3}{ab}}{5b + \dfrac{2}{ab}}$$

Solution: First simplify the numerator and the denominator of the complex fraction.

$$\dfrac{2a - \dfrac{3}{ab}}{5b + \dfrac{2}{ab}} = \dfrac{\dfrac{2a \cdot ab}{ab} - \dfrac{3}{ab}}{\dfrac{5b \cdot ab}{ab} + \dfrac{2}{ab}}$$ The LCD is ab.

The LCD is ab.

$$= \dfrac{\dfrac{2a^2b - 3}{ab}}{\dfrac{5ab^2 + 2}{ab}}$$ Subtract.

Add.

$$= \frac{2a^2b - 3}{ab} \cdot \frac{ab}{5ab^2 + 2}$$

Multiply by the reciprocal of $\dfrac{5ab^2 + 2}{ab}$.

$$= \frac{ab(2a^2b - 3)}{ab(5ab^2 + 2)}$$

$$= \frac{2a^2b - 3}{5ab^2 + 2}$$

Simplify.

EXAMPLE 2   *Objective 3: Simplify complex fractions by multiplying by a common denominator.*

Simplify the complex fraction.

$$\frac{2a - \dfrac{3}{ab}}{5b + \dfrac{2}{ab}}$$

Solution: The least common denominator of $2a - \dfrac{3}{ab}$ and $5b + \dfrac{2}{ab}$ is ab. Multiply both the numerator and the denominator of the complex fraction by the LCD.

$$\frac{2a - \dfrac{3}{ab}}{5b + \dfrac{2}{ab}} = \frac{\left(2a - \dfrac{3}{ab}\right) \cdot ab}{\left(5b + \dfrac{2}{ab}\right) \cdot ab}$$

Multiply numerator and denominator by the LCD.

$$= \frac{2a \cdot ab - \dfrac{3}{ab} \cdot ab}{5b \cdot ab + \dfrac{2}{ab} \cdot ab}$$

Apply the distributive property.

$$= \frac{2a^2b - 3}{5ab^2 + 2}$$

Simplify.

EXAMPLE 3   *Objective 4: Simplify expressions with negative exponents.*

Simplify.

$$\frac{2x^{-2} - x^{-2}y}{4x^{-1}(1 - y) + x^{-1}y^2}$$

Solution: This fraction does not appear to be a complex fraction. If we write it using only positive exponents, however, we see that it is a complex fraction.

$$\frac{2x^{-2} - x^{-2}y}{4x^{-1}(1 - y) + x^{-1}y^2} = \frac{\dfrac{2}{x^2} - \dfrac{y}{x^2}}{\dfrac{4}{x} - \dfrac{4y}{x} + \dfrac{y^2}{x}}$$

Write with positive exponents only.

$$= \frac{\dfrac{2 - y}{x^2}}{\dfrac{y^2 - 4y + 4}{x}}$$

Combine like terms.

$$= \frac{2 - y}{x^2} \cdot \frac{x}{y^2 - 4y + 4}$$

Multiply by the reciprocal of $\dfrac{y^2 - 4y + 4}{x}$.

$$= \frac{(-1) \cdot x \cdot (y-2)}{x^2 \cdot (y-2) \cdot (y-2)} \qquad \text{Factor.}$$

$$= -\frac{1}{x(y-2)} \qquad \text{Simplify.}$$

EXERCISES

Exercises 1–6. Simplify each complex fraction. See Examples 1 and 2.

1. $\dfrac{\dfrac{3x}{2}}{\dfrac{x^2}{4}}$
  2. $\dfrac{\dfrac{3}{y}}{\dfrac{9}{2y}}$
  3. $\dfrac{2+\dfrac{3x}{2}}{4-\dfrac{5x}{3}}$
  4. $\dfrac{1-\dfrac{5}{4x}}{2+\dfrac{3}{x}}$

5. $\dfrac{\dfrac{5x}{4x^2-9}}{\dfrac{7x^2-3x}{2x+3}}$
  6. $\dfrac{1+\dfrac{1}{w+3}}{1+\dfrac{7}{w-3}}$

Exercises 7–10. Simplify. See Example 3.

7. $\dfrac{y^{-2}}{y^{-1}+(2y)^{-1}}$
  8. $\dfrac{x^2}{x^{-1}+x^{-2}}$
  9. $\dfrac{5-w^{-2}}{1+w^{-1}}$
  10. $\dfrac{3q^{-2}+4r^{-1}}{2q^{-1}+r^{-2}}$

Exercises 11–12. In the study of calculus, the difference quotient $\dfrac{f(a+h)-f(a)}{h}$ *is often found and simplified.*
Find and simplify this quotient for each function f(x) below by following steps (a) through (d).
(a) *Find f(a + h).*
(b) *Find f(a).*
(c) *Use steps (a) and (b) to find* $\dfrac{f(a+h)-f(a)}{h}$.
(d) *Simplify the result of step (c).*

11. $f(x) = \dfrac{1}{x-2}$
  12. $f(x) = \dfrac{x}{x+3}$

C *Exercises 13–14. Simplify each complex fraction. Then use a grapher to confirm your work by graphing the original fraction and your simplified answer on the same screen.*

13. $\dfrac{\dfrac{2x-1}{3}}{\dfrac{x+3}{9}}$
  14. $\dfrac{\dfrac{3}{x^2}-\dfrac{1}{x}}{\dfrac{2}{x}+5}$

The following circle graphs represent the percent of new car sales by type in the United States for 1984 and 1994. Use these graphs to answer Exercises 15–19. (data from American Automobile Manufacturers Association)

1984

1994

15. In 1984, midsize cars accounted for what percent of car sales?
16. In 1994, small cars accounted for what percent of car sales?
17. Which category experienced the largest increase from 1984 to 1994? By what percent?
18. Which category experienced the largest decrease from 1984 to 1994? By what percent?
19. Did sales of luxury cars increase or decrease from 1984 to 1994? By what percent?

6.5   Dividing Polynomials

EXAMPLE 1   *Objective 1: Divide by a monomial.*

Find the quotient: $\dfrac{8w^2z - 2wz + 4wz^2}{2wz}$.

Solution: Divide each term of the polynomial in the numerator by $2wz$ and simplify.

$$\frac{8w^2z - 2wz + 4wz^2}{2wz} = \frac{8w^2z}{2wz} - \frac{2wz}{2wz} + \frac{4wz^2}{2wz} = 4w - 1 + 2z$$

EXAMPLE 2   *Objective 2: Divide by a polynomial.*

Divide $6x^4 + 9x^3 + 2x^2 - 8x - 3$ by $3x^2 + 1$.

Solution: Before dividing, we write any "missing powers" as the product of 0 and the variable raised to the missing power. There is no x term in the divisor, so include $0x$ in the divisor.

$$3x^2 + 0x + 1 \overline{) 6x^4 + 9x^3 + 2x^2 - 8x - 3} \qquad \dfrac{2x^2 + 3x + 0}{}$$

$\dfrac{6x^4}{3x^2} = 2x^2$.

$$\underline{6x^4 + 0x^3 + 2x^2} \qquad\qquad 2x^2(3x^2 + 0x + 1) .$$
$$0x^4 + 9x^3 + 0x^2 - 8x \qquad\qquad \text{Subtract. Bring down } -8x.$$
$$\underline{9x^3 + 0x^2 + 3x} \qquad\qquad 3x(3x^2 + 0x + 1) .$$
$$-11x - 3 \qquad\qquad \text{Subtract. Bring down } -3.$$

At this stage the degree of the remainder is less than the degree of the divisor, so there is no constant term in the quotient (we use a 0 as a place holder in the quotient). Thus,

$$\frac{6x^4 + 9x^3 + 2x^2 - 8x - 3}{3x^2 + 1} = 2x^2 + 3x - \frac{11x + 3}{3x^2 + 1}$$

To check, see that $6x^4 + 9x^3 + 2x^2 - 8x - 3 = (3x^2 + 1)(2x^2 + 3x) - (11x + 3)$.

EXERCISES

Exercises 1–4. Find each quotient. See Example 1.

1. $\dfrac{90x^2 + 50x - 10}{10x}$

2. $\dfrac{25a^3 + 15a^2 - 45a}{5a^3}$

3. $\dfrac{56q^2r^2 - 24q^2r + 28qr^2 + 4qr}{4q^2r^2}$

4. $\dfrac{12xy^2 + 10x^2y - 18xy}{6xy^3}$

Exercises 5–8. Find each quotient. See Example 2.

5. $\dfrac{2y^2 + 9y - 18}{y + 6}$

6. $\dfrac{x^3 + 3x^2 - 2x - 6}{x^2 - 2}$

7. $\dfrac{3w^3 + 12w^2 - w - 2}{3w^2 - 1}$

8. $\dfrac{6x^4 - 2x^3 + 3x + 5}{3x - 1}$

9. The perimeter of a regular pentagon is $15x^3 - 3x^2 + 40x + 85$ meters. Find the length of each side.

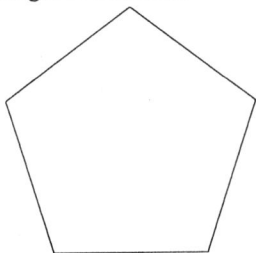

10. The area of a parallelogram is $8x^2 - 2x - 1$ square inches, and its base is $4x + 1$ inches. Find the height of the parallelogram.

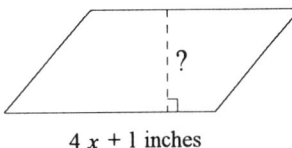

$4x + 1$ inches

11. A race walker walks for $2x - 1$ hours and covers a distance of $4x^2 + 12x - 7$ miles. Find the speed of the walker in miles per hour.

12. The formula for the area of a trapezoid is $A = \frac{1}{2}h(b_1 + b_2)$, where h represents the height of the trapezoid and b_1 and b_2 represent the lengths of the two parallel sides, or bases. If the area of a trapezoid is $6x^2 - 7x - 3$ square centimeters and the lengths of the two bases are $2x + 6$ centimeters and $4x - 4$ centimeters, find the height of the trapezoid.

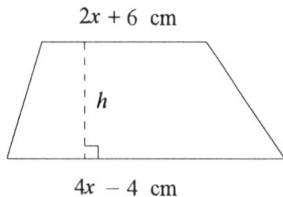

C *Exercises 13–14. Find each quotient. Then use a grapher to confirm your work by graphing the original fraction and your simplified answer on the same screen.*

13. $\dfrac{3x^2 - 10x - 25}{3x + 5}$

14. $\dfrac{x^2 + 5x + 3}{x + 3}$

6.6   Synthetic Division and the Remainder Theorem

EXAMPLE 1   *Objective 1: Use synthetic division to divide a polynomial by a binomial.*

Use synthetic division to divide $6x^4 + 9x^3 - 20x^2 + 11x - 3$ by $x + 3$.

Solution: The divisor is $x + 3$, which we write in the form $x - c$ as $x - (-3)$. Thus, c is -3. The dividend coefficients are 6, 9, –20, 11, and –3.

$$
\begin{array}{r|rrrrr}
-3 & 6 & 9 & -20 & 11 & -3 \\
   &   & -18 & 27 & -21 & 30 \\
\hline
   & 6 & -9 & 7 & -10 & 27
\end{array}
$$

The dividend is a fourth-degree polynomial, so the quotient polynomial is a third-degree polynomial. The quotient is $6x^3 - 9x^2 + 7x - 10$ with a remainder of 27. Thus,

$$\frac{6x^4 + 9x^3 - 20x^2 + 11x - 3}{x + 3} = 6x^3 - 9x^2 + 7x - 10 + \frac{27}{x + 3}$$

EXAMPLE 2   *Objective 2: Use the remainder theorem to evaluate polynomials.*

Use the remainder theorem and synthetic division to find $P(-3)$ if
$$P(x) = 6x^4 + 9x^3 - 20x^2 + 11x - 3$$

Solution: To find $P(-3)$ by the remainder theorem, we use synthetic division to divide $P(x)$ by $x + 3$. In Example 2, we saw that $6x^4 + 9x^3 - 20x^2 + 11x - 3$ divided by $x + 3$ resulted in a remainder of 27. Thus, $P(-3) = 27$, the remainder. Use a calculator to check this result by evaluating $P(x)$ for $x = -3$.

EXERCISES

Exercises 1–6. Use synthetic division to find each quotient. See Example 1.

1. $\dfrac{9x^2 + 5x - 7}{x - 1}$

2. $\dfrac{a^3 - 15a^2 - 100}{a - 5}$

3. $\dfrac{4q^2 + 19q - 5}{q + 5}$

4. $\dfrac{2y^2 + 9y - 18}{y + 6}$

5. $\dfrac{6x^4 - 2x^3 + 3x + 5}{x - 1}$

6. $\dfrac{2x^5 - 5x^4 - 3x^2 - 6x - 10}{x - 3}$

Exercises 7–10. For the given polynomial P(x) and the given value of c, find P(c) using the remainder theorem. See Example 2.

7. $P(x) = 3x^3 + 12x^2 - x - 2$; -4

8. $P(x) = 8x^4 - 5x^2 + 46$; 5

9. $P(x) = x^5 + x^4 + 500$; -5

10. $P(x) = -x^3 + 5x^2 - 3x + 7$; 3

11. If the volume of a box is $15y^3 + 354y^2 - 144y$ cubic centimeters, its width is $3y$ centimeters, and its length is $y + 24$ centimeters, find its height.

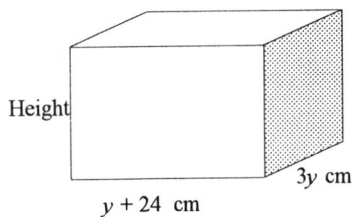

Height
$y + 24$ cm
$3y$ cm

12. If the area of a parallelogram is $2x^3 - 11x^2 - 42x + 147$ square feet and its base is $x - 7$ feet, find its height.

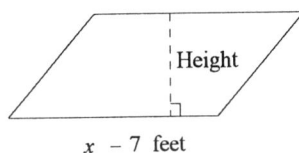

Height
$x - 7$ feet

6.7   Solving Equations Containing Rational Expressions

EXAMPLE 1   *Objective 1: Solve equations containing rational expressions.*

Solve: $\dfrac{x^2+2x+4}{x^2+6x+8}+\dfrac{x}{x+2}=\dfrac{x-3}{x+4}$.

Solution: Factor the first denominator to find that the LCD is $(x+4)(x+2)$. Multiply both sides of the equation by $(x+4)(x+2)$. By the distributive property, this is the same as multiplying each term by $(x+4)(x+2)$.

$$\frac{x^2+2x+4}{x^2+6x+8}+\frac{x}{x+2}=\frac{x-3}{x+4}$$

$$(x+4)(x+2)\frac{x^2+2x+4}{x^2+6x+8}+(x+4)(x+2)\frac{x}{x+2}=(x+4)(x+2)\frac{x-3}{x+4}$$

$$x^2+2x+4+x(x+4)=(x+2)(x-3)$$

$$x^2+2x+4+x^2+4x=x^2-x-6$$

$$2x^2+6x+4=x^2-x-6$$

$$x^2+7x+10=0$$

$$(x+5)(x+2)=0$$

$$x=-5 \quad \text{or} \quad x=-2$$

Now by checking each proposed solution in the original equation, we see that -2 makes the denominator of the term $\dfrac{x}{x+2}$ equal to zero. Thus -2 is an extraneous solution, and the solution set is $\{5\}$.

EXERCISES

Exercises 1–8. Solve each equation. See Example 1.

1. $\dfrac{a}{4}-\dfrac{a}{6}=\dfrac{2}{3}$

2. $\dfrac{y+6}{5}=\dfrac{y+8}{3}$

3. $\dfrac{2x-7}{6x+5}=\dfrac{x-4}{3x-1}$

4. $\dfrac{-2x}{3x+2}=\dfrac{3x-5}{3x+2}$

5. $\dfrac{21}{(y+3)^2}=2-\dfrac{1}{y+3}$

6. $\dfrac{5}{3x+6}+\dfrac{11}{9}=\dfrac{4}{x+2}$

7. $\dfrac{3x}{x-3}-\dfrac{18}{x^2-4x+3}=\dfrac{11}{x-1}$

8. $\dfrac{x}{3x-1}=\dfrac{1}{9x+6}+\dfrac{1}{9x^2+3x-2}$

9. The average cost of producing x graphing calculators is given by the function $f(x) = 17.50 + \dfrac{5000}{x}$. Find the number of graphing calculators that must be produced for the average cost to be $20.

10. The average cost of producing x boxes of cereal is given by the function $f(x) = 1.50 + \dfrac{1000}{x}$. Find the number of boxes of cereal that must be produced for the average cost to be $2.75.

C *Exercises 11–14. Use a grapher to verify the solution of each given exercise.*

11. Exercise 5

12. Exercise 7

13. Exercise 8

14. Exercise 9

6.8 Rational Equations and Problem Solving

EXAMPLE 1 *Objective 2: Solve an equation containing rational expressions for a specified variable.*

An elevated water reservoir supplies water to two nearby towns. One town can empty the reservoir in 10 hours by itself. The other town can empty the reservoir in 15 hours by itself. It takes 20 hours for an inlet pipe to completely fill the reservoir. If the reservoir is completely filled and the two outlets for the towns are opened at the same time as the inlet pipe for filling the reservoir, how long will it take to completely drain the reservoir?

Solution: 1. UNDERSTAND. Read and reread the problem. The key idea here is the relationship between the times (in hours) it takes each town to empty the reservoir and the time it takes to fill the reservoir. If it takes the first town 10 hours to empty the reservoir, the part of the reservoir it can empty in 1 hour is $\dfrac{1}{10}$. Similarly, if it takes the second town 15 hours to empty the reservoir, the part of the reservoir it can empty in 1 hour is $\dfrac{1}{15}$. Also, if it takes 20 hours to fill the reservoir, the part of the reservoir it can *fill* in 1 hour is $\dfrac{1}{20}$. Later, we must be careful to consider filling the reservoir as an action *opposite* to draining the reservoir.

2. ASSIGN. Let t represent the time in hours it takes to completely drain the reservoir while it is being drained by both towns and filled simultaneously. Then $\dfrac{1}{t}$ represents the part of the reservoir that has been drained in 1 hour.

3. ILLUSTRATE. Here we summarize the information discussed above on a chart.

	Hours to complete the job	Part of job completed in 1 hour
First town to drain the reservoir	10	$\frac{1}{10}$
Second town to drain the reservoir	15	$\frac{1}{15}$
Inlet pipe to fill the reservoir	20	$\frac{1}{20}$
Together	t	$\frac{1}{t}$

4. TRANSLATE.

In words:	part of reservoir drained by 1st town in 1 hour	added to	part of reservoir drained by 2nd town in 1 hour	decreased by	part of reservoir *filled* by inlet pipe in 1 hour	is equal to	overall part of reservoir drained in 1 hour
Translate:	$\frac{1}{10}$	+	$\frac{1}{15}$	−	$\frac{1}{20}$	=	$\frac{1}{t}$

5. COMPLETE.

$$\frac{1}{10}+\frac{1}{15}-\frac{1}{20}=\frac{1}{t}$$
$$60t\left(\frac{1}{10}+\frac{1}{15}-\frac{1}{20}\right)=60t\left(\frac{1}{t}\right) \qquad \text{Multiply both sides by the LCD } 60t.$$
$$6t+4t-3t=60$$
$$7t=60$$
$$t=\frac{60}{7} \text{ or } 8\frac{4}{7}$$

6. INTERPRET. *Check:* The proposed solution is $8\frac{4}{7}$. That is, with the 2 towns draining the reservoir at the same time it is being filled, it will take $8\frac{4}{7}$ hours to completely drain the reservoir. Check this solution in the originally stated problem.

State: It will take $8\frac{4}{7}$ hours to completely drain the reservoir if the 2 towns are draining the reservoir at the same time that it is being filled.

EXERCISES

1. The sum of a number and 9 times its reciprocal is 6. Find the number(s).

2. The difference of a number and 21 times its reciprocal is 4. Find the number(s).

3. The quotient of a number and twice its reciprocal is 8. Find the number(s).

4. The quotient of a number and 9 times its reciprocal is $\dfrac{1}{25}$. Find the number(s).

5. The speed of a current is 8 mph. If it takes a boat the same time to travel 12 miles upstream as it takes to travel 24 miles downstream, find the speed of the boat in still water.

6. The speed of a person on in-line roller skates is 4 mph slower than the speed of a bicyclist. If the in-line skater can travel 5 miles in the same amount of time as the bicyclist can travel 8 miles, find the speed of the in-line skater.

7. A parts distributor's warehouse can be completely filled in 12 hours. Shipments to customers can empty the filled warehouse in 6 hours. If the warehouse is completely full and shipments into the warehouse and shipments out to customers occur at the same time, how long will it take to empty the warehouse?

8. One person can rake a yard in 4 hours. If a second person helps with the raking, it takes only 3 hours to rake the yard. How long would it take the second person to rake the yard working alone?

9. Normally, an above-ground swimming pool can be filled with a hose in 6 hours. The pool developed a leak that drained the pool in 36 hours. How long will it take to fill the pool if the leak is not repaired first?

10. A fruit punch mixing vat can be emptied in 2 hours by dispensing fruit punch into containers on a conveyor belt. Water is added to the vat through valve #1, which can fill the vat in 3 hours. Fruit punch concentrate is added to the vat through valve #2, which can fill the vat in 7 hours. Assume that the vat is currently full. How long will it take to empty the vat if both valves are opened at the same time as fruit punch begins to be dispensed?

6.9   Variation and Problem Solving

EXAMPLE 1   *Objective 1: Write an equation expressing direct variation.*

You are paid an hourly wage. Suppose that for one week you worked 45 hours and were paid $562.50. What will you earn if you work for 48.5 hours during one week?

Solution: 1. UNDERSTAND. Read and reread the problem. Notice that being paid an hourly wage implies that your weekly pay is directly proportional to the number of hours you work during the week.

2. ASSIGN. Let y represent your weekly pay and let x represent the number of hours you worked during the week.

4. TRANSLATE. Because y is directly proportional to x, we write
$$y = kx$$

When you work 45 hours, you are paid $562.50. That is, when $x = 45$, $y = 562.50$.

$$y = kx$$
$$562.50 = k(45) \qquad \text{Replace by the known values.}$$
$$12.50 = k \qquad \text{Solve for } k.$$

Thus, $y = 12.50x$. To find your pay when you work 48.5 hours during one week, replace x with 48.5 and find y.

5. COMPLETE.
$$y = 12.50x$$
$$y = 12.50(48.5) = 606.25$$

6. INTERPRET. *Check* the proposed solution of $606.25. *State:* You will earn $606.25 during a week in which you work 48.5 hours.

EXAMPLE 2   *Objective 2: Write an equation expressing inverse variation.*

The sales of hot chocolate at a football game are inversely proportional to the temperature during half time. If 50 cups of hot chocolate are sold when the temperature is 55° Fahrenheit during half time, find how many cups of hot chocolate will be sold when the temperature is 35° Fahrenheit during half time.

Solution: 1. UNDERSTAND. Read and reread the problem. Notice that we are given that the sales of hot chocolate are **inversely proportional** to the temperature at half time.

2. ASSIGN. Let y represent the number of cups of hot chocolate sold and let x represent the temperature during half time.

4. TRANSLATE. Because y is inversely proportional to x, we write
$$y = \frac{k}{x}$$

When the temperature is 55 degrees, 50 cups of hot chocolate are sold. That is, when $x = 55$, $y = 50$.

$$y = \frac{k}{x}$$

$$50 = \frac{k}{55} \qquad \text{Replace by the known values.}$$

$$50 \cdot 55 = k \qquad \text{Solve for } k.$$

$$2750 = k \qquad \text{Simplify.}$$

Thus, $y = \dfrac{2750}{x}$. To find the number of cups of hot chocolate sold when the temperature is 35 degrees, replace x with 35 and find y.

5. COMPLETE.

$$y = \frac{2750}{x}$$

$$y = \frac{2750}{35} \approx 79$$

6. INTERPRET. *Check* the proposed solution of 79 cups of hot chocolate. *State:* Approximately 79 cups of hot chocolate will be sold when the temperature at half time is 35° Fahrenheit.

EXAMPLE 3   *Objective 3: Write an equation expressing joint variation.*

Suppose that Q varies jointly with r and the square of p. Write a formula relating these variables.

Solution: $Q = kp^2 r$

EXERCISES

Exercises 1–4. Write each statement as an equation.

1. P varies inversely as q.

2. W is directly proportional to z.

3. G varies jointly as t and s.

4. B is jointly proportional to x and the square of y.

Exercises 5–10. Solve. See Examples 1–3.

5. If the voltage V in an electric circuit is held constant, the current I is inversely proportional to the resistance R. If the current is 32 amperes when the resistance is 25 ohms, find the current when the resistance is 20 ohms.

6. At an amusement park, the daily sales of ice cream are directly proportional to the square of the daily high temperature. If 3698 ice cream cones are sold on a day when the high temperature is 86° Fahrenheit, how many ice cream cones will be sold when it is 75° Fahrenheit?

7. The number of hours it takes a math professor to grade final exams varies jointly as the number of students in each class and the number of classes the professor teaches. If it takes the professor 72 hours to grade final exams when there are 32 students in each of the 3 classes the professor teaches, how long will it take to grade final exams when the professor teaches 4 classes of 21 students each?

8. The weight of an object on the moon is directly proportional to its weight on the earth. A person weighing 150 pounds on earth would weigh 25.5 pounds on the moon. How much would a tool pack weighing 80 pounds on earth weigh on the moon?

9. The magnification power of a microscope is inversely proportional to the wavelength of the light used in the magnification process. If light with a wavelength of 4000 angstroms produces a magnification of 125 times, what magnification power does light with a wavelength of 6000 angstroms produce?

10. The area of an ellipse varies jointly as the length of its major axis a and the length of its minor axis b. If the area of an ellipse with $a = 28$ centimeters and $b = 18$ centimeters is 126π square centimeters, what is the area of an ellipse with $a = 54$ and $b = 16$?

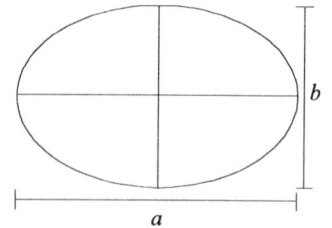

The bar graph below shows the most popular colors for compact and sports car for the 1994 model year. Use the graph to answer the questions in Exercises 11–16. (data from American Automobile Manufacturers Association)

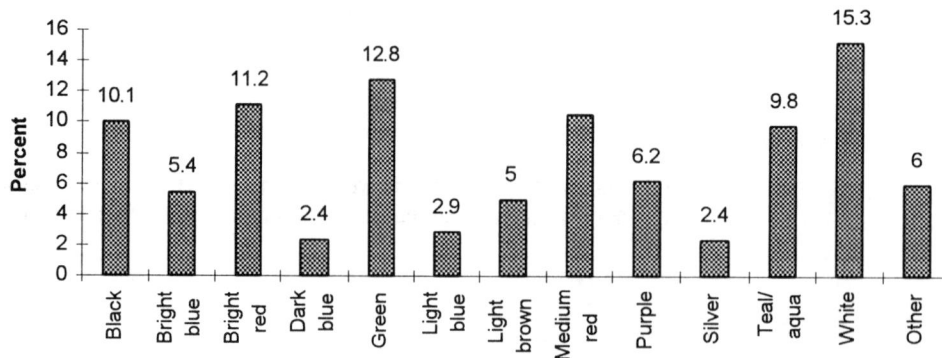

11. What percent of compact sports cars were purple?
12. What percent of compact and sports cars were medium red?
13. What car color was the most popular for compact and sports cars?
14. What color was the least popular?
15. What percent of compact and sports cars were either bright or medium red?
16. In 1994, the United States produced 6,613,970 passenger cars. Of these, 29.2% were compact/sports cars. Approximately how many compact/sports cars were black in 1994? (data from American Automobile Manufacturers Association)

Chapter 6   Hints and Warnings

- To **simplify a rational expression:**
    Step 1: Completely factor the numerator and the denominator. *Don't forget that if you have a factor of* $(a-x)$*, where a is any constant, in either the numerator or denominator, you can remove a factor of* -1 *out to get* $-1(x-a)$*. This can be helpful especially if there are other factors of* $(x-a)$ *elsewhere in the rational expression.*
    Step 2: Apply the fundamental principle of rational expressions to divide both the numerator and denominator by their GCF.
- To **multiply rational expressions:**
    Step 1: Completely factor the numerators and denominators.
    Step 2: Multiply the numerators and multiply the denominators.
    Step 3: Apply the fundamental principle of rational expressions.
- To **divide rational expressions,** multiply the first rational expression by the reciprocal of the second rational expression and proceed with the step for multiplying rational expressions. *Don't forget that the reciprocal of* $\dfrac{1}{a}$ *is a.*
- To **add or subtract rational expressions:**
    Step 1: Find the LCD.
    Step 2: Write each rational expression as an equivalent rational expression whose denominator is the LCD.
    Step 3: Add or subtract numerators and write the sum or difference over the common denominator.
    Step 4: Write the result in lowest terms.
- To **divide a polynomial by a monomial,** divide each term in the polynomial be the monomial. *Don't forget to write each resulting term of the polynomial in lowest terms.*
- To **divide a polynomial by a polynomial other than a monomial,** use long division. *Remember how long division of polynomials parallels long division of real numbers.*
- To **solve an equation containing rational expressions,** multiply both sides of the equation by the LCD of all rational expressions in the equation. Then apply the distributive property as appropriate and simplify. Solve the resulting equation, and then check the proposed solution to see whether it makes the denominator 0. If so, it is an extraneous solution. Exclude any extraneous solutions from the solution set. *Once fractions have been eliminated from the equation by multiplying by the LCD, you can solve the resulting equation (either linear or quadratic, etc.) using the methods you have learned previously for solving such equations.*
- To **solve an equation for a specified variable,** treat the specified variable as the only variable in the equation and solve as usual.
- To **solve problems involving variation,** remember that the phrase "y varies directly as x" or "y is directly proportional to x" can be translated as $y = kx$, and the phrase "y varies inversely as x" or "y is inversely proportional to x" can be translated as $y = \dfrac{k}{x}$.

CHAPTER 6 Practice Test

Find the domain of each rational function.

1. $f(x) = \dfrac{6x^2}{5x - 15}$

2. $f(x) = \dfrac{2x + 5}{x^2 + 10x + 25}$

Write each rational expression in lowest terms.

3. $\dfrac{6 - 2x}{5x - 15}$

4. $\dfrac{3x^2 + 9x}{6x^3 - 9x^2 + 15x}$

Perform the indicated operation.

5. $\dfrac{y}{y + 3} \cdot \dfrac{y^2 - 9}{8y}$

6. $\dfrac{38qr^2}{12p} \div \dfrac{19q^3r}{3p^4qr^2}$

7. $\dfrac{6x}{x + 2} + \dfrac{10}{x - 7}$

8. $\dfrac{9x}{x^2 + 4x + 3} - \dfrac{5}{x + 3}$

Divide. Simplify each answer.

9. $\dfrac{\dfrac{2}{w} + \dfrac{3}{2w}}{\dfrac{5}{3w} - \dfrac{1}{w}}$

10. $\dfrac{18a^2b^2c + 9ab - 6bc^2}{3abc}$

11. $\dfrac{x^3 + 5x^2 - 9x - 45}{x + 5}$

12. Use synthetic division to divide $3x^4 - 3x^3 + 2x^2 + 20x - 11$ by $x + 2$.

13. If $f(x) = 4x$ and $g(x) = 10 - 7x$, find $(f + g)(x)$ and $(g \cdot f)(x)$.

Solve each equation for x.

14. $\dfrac{x - 2}{4x + 1} = \dfrac{15}{4}$

15. $\dfrac{x}{x + 4} - \dfrac{2}{x - 3} = 1$

16. $\dfrac{8x}{x + 5} = 12 - \dfrac{4}{x + 5}$

17. The product of three less than a number and four times the reciprocal of the number is $\dfrac{16}{7}$. Find the number.

18. If one person can do a job in 10 hours and a second person can do the same job in 14 hours, how long will it take them to do the job working together?

19. Suppose that R is inversely proportional to w. If $R = 120$ and $w = 8$, find w when $R = 200$.

20. Suppose the P is jointly proportional to t and s. If $P = 60$ when $t = 5$ and $s = 2$, find P when $t = 20$ and $s = 51$.

CHAPTER 7

Rational Exponents, Radicals, and Complex Numbers

7.1   Radicals and Radical Functions

EXAMPLE 1   *Objective 3: Find nth roots.*

Simplify the following expressions. Assume that variables represent positive real numbers.

a. $\sqrt[6]{64x^{12}z^6}$ b. $\sqrt[5]{-32a^5b^{15}}$ c. $\sqrt[4]{(-7)^4}$

Solution: a. $\sqrt[6]{64x^{12}z^6} = 2x^2z$ because $(2x^2z)^6 = 64x^{12}z^6$.

b. $\sqrt[5]{-32a^5b^{15}} = -2ab^3$ because $(-2ab^3)^5 = -32a^5b^{15}$.

c. $\sqrt[4]{(-7)^4} = |-7| = 7$

EXAMPLE 2   *Objective 4: Graph square root functions.*

Graph the function $f(x) = \sqrt{x-2} + 2$.

Solution: To graph, we identify the domain, evaluate the function for several values of x, plot the resulting points, and connect the points with a smooth curve. The domain of this function is the set of all satisfying $x - 2 \geq 0$ or $\{x | x \geq 2\}$.

x	$f(x) = \sqrt{x-2} + 2$
2	$\sqrt{2-2} + 2 = 2$
3	$\sqrt{3-2} + 2 = 3$
4	$\sqrt{4-2} + 2 = \sqrt{2} + 2 \approx 3.41$
6	$\sqrt{6-2} + 2 = 4$
11	$\sqrt{11-2} + 2 = 5$

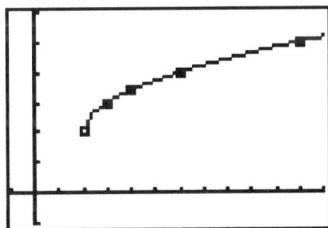

Notice that the graph of this function passes the vertical line test as expected.

EXAMPLE 3   *Objective 4: Graph cube root functions.*

Graph the function $f(x) = \sqrt[3]{x+4} - 3$.

Solution: To graph this function, we identify the domain, plot points and connect the points with a smooth curve. The domain of this function is the set of all real numbers.

x	$f(x) = \sqrt[3]{x+4} - 3$
-12	$\sqrt[3]{-12+4} - 3 = -5$
-7	$\sqrt[3]{-7+4} - 3 = \sqrt[3]{-3} - 3 \approx -4.44$
-5	$\sqrt[3]{-5+4} - 3 = -4$
-4	$\sqrt[3]{-4+4} - 3 = -3$
-3	$\sqrt[3]{-3+4} - 3 = -2$
-1	$\sqrt[3]{-1+4} - 3 = \sqrt[3]{3} - 3 \approx -1.56$
0	$\sqrt[3]{0+4} - 3 = \sqrt[3]{4} - 3 \approx -1.41$
4	$\sqrt[3]{4+4} - 3 = -1$

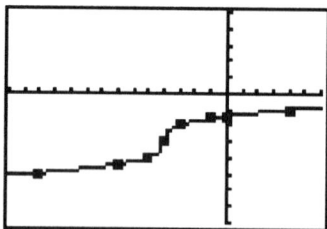

EXERCISES

Exercises 1–8. Simplify each radical. Assume that all variables represent positive real numbers. See Example 1.

1. $\sqrt{1600}$

2. $\sqrt[3]{-15{,}625}$

3. $\sqrt{49x^6y^4}$

4. $\sqrt[3]{27a^3b^9}$

5. $\sqrt[5]{p^{15}q^5r^{25}}$

6. $\sqrt[5]{-t^5}$

7. $\sqrt[4]{(q-4)^4}$

8. $\sqrt[6]{-36}$

Exercises 9–12. Identify the domain and then graph each function. See Examples 2 and 3.

9. $f(x) = \sqrt{x} - 4$

x	$f(x)$

10. $f(x) = \sqrt{x+3}$

x	$f(x)$
-3	
0	
1	
6	

11. $f(x) = \sqrt[3]{x-2}$

x	$f(x)$
-6	
-1	
3	
10	

12. $f(x) = \sqrt[3]{x} + 3$

x	$f(x)$

C *Exercises 13–14. Use a grapher to graph and find the domain of each function.*

13. $f(x) = \sqrt{x-7} + 6$

14. $f(x) = \sqrt[3]{-x+5} - 1$

7.2   Rational Exponents

EXAMPLE 1   *Objectives 2 and 3: Understand the meanings of a raised to the $\frac{m}{n}$ th power and a raised to the $-\frac{m}{n}$ th power.*

Simplify using radical notation.

a. $9^{5/2}$ b. $8^{-2/3}$

Solution: a. $9^{5/2} = \left(\sqrt{9}\right)^5 = 3^5 = 243$

b. $8^{-2/3} = \dfrac{1}{8^{2/3}} = \dfrac{1}{\sqrt[3]{8^2}} = \dfrac{1}{\sqrt[3]{64}} = \dfrac{1}{4}$

EXAMPLE 2   *Objective 4: Use the rules of exponents to simplify expressions containing radicals.*

Write using rational exponents. Then simplify if possible.

a. $\sqrt[3]{-8x^6yz^4}$ b. $\sqrt[3]{x^4}\sqrt{x}$

Solution: a. $\sqrt[3]{-8x^6yz^4} = (-8x^6yz^4)^{1/3} = (-8)^{1/3}(x^6)^{1/3}y^{1/3}(z^4)^{1/3} = -2x^{6/3}y^{1/3}z^{4/3} = -2x^2y^{1/3}z^{4/3}$

b. $\sqrt[3]{x^4}\sqrt{x} = x^{4/3}x^{1/2} = x^{(4/3+1/2)} = x^{(8/6+3/6)} = x^{11/6}$

EXAMPLE 3   *Objective 4: Use the rules of exponents to simplify expressions containing radicals.*

Write using properties of exponents.

a. $a^{1/4}(a^{1/2} + 3a^4)$ b. $\left(\dfrac{x}{x^{3/2}}\right)^{2/3}$

Solution: a. $a^{1/4}(a^{1/2} + 3a^4) = a^{1/4}a^{1/2} + a^{1/4} \cdot 3a^4$

$= a^{(1/4+1/2)} + 3a^{1/4+4}$

$= a^{3/4} + 3a^{17/4}$

b. $\left(\dfrac{x}{x^{3/2}}\right)^{2/3} = \left(x^{1-3/2}\right)^{2/3} = \left(x^{-1/2}\right)^{2/3} = x^{-1/3} = \dfrac{1}{x^{1/3}}$

EXERCISES

Exercises 1–4. Write the following using radical notation. Simplify if possible. See Example 1.

1. $(9m)^{3/2}$ **2.** $(-8)^{1/3}$ **3.** $5a^{4/5}$ **4.** $(x+7)^{2/3}$

Exercises 5–8. Write with positive exponents. Simplify if possible. See Example 1.

5. $x^{-3/4}$ **6.** $(-27)^{2/3}$ **7.** $\dfrac{2}{5m^{-5/2}}$ **8.** $3q^{-1/7}$

Exercises 9–12. Write using rational exponents. Then simplify if possible. See Example 2.

9. $\sqrt[7]{q^2}$ **10.** $\sqrt[5]{4r^3}$ **11.** $\sqrt{(3r-10)^4}$ **12.** $4\sqrt{a} - 5\sqrt[3]{b}$

Exercises 13–16. Use properties of exponents to simplify each expression. See Example 3.

13. $q^{3/5}q^{7/5}$ **14.** $\dfrac{a^{1/3}b^{3/4}}{a^{2/3}}$ **15.** $\left(x^2y^6z^3\right)^{1/2}$ **16.** $\left(p^{1/3}q^{1/4}r^{4/3}\right)^3$

Exercises 17–18. Factor the common term from the expression.

17. $x^{2/5}$; $2x^{4/5} - 3x^{2/5}$ **18.** $5x^{4/3}$; $5x^{7/3} + 10x^{4/3}$

C *Exercises 19–20. Use a calculator to write a four decimal-place approximation for each.*

19. $14^{2/3}$ **20.** $25^{5/4}$

7.3   Simplifying Radical Expressions

EXAMPLE 1   *Objectives 1 and 2: Understand the product and quotient rules for radicals.*

Perform the indicated operation.

     a. $\sqrt[3]{5x} \cdot \sqrt[3]{10xy^2}$ **b.** $\dfrac{\sqrt[4]{20}}{\sqrt[4]{2}}$

Solution: **a.** $\sqrt[3]{5x} \cdot \sqrt[3]{10xy^2} = \sqrt[3]{5x \cdot 10xy^2} = \sqrt[3]{50x^2y^2}$

            b. $\dfrac{\sqrt[4]{20}}{\sqrt[4]{2}} = \sqrt[4]{\dfrac{20}{2}} = \sqrt[4]{10}$

EXAMPLE 2   *Objective 3: Simplify radicals.*

Simplify the following.

     a. $\sqrt{45}$ **b.** $\sqrt[3]{48a^4b^5c^6}$ **c.** $\dfrac{\sqrt[4]{160x^7}}{\sqrt[4]{10x^3}}$

Solution: **a.** $\sqrt{45} = \sqrt{9 \cdot 5} = \sqrt{9} \cdot \sqrt{5} = 3\sqrt{5}$

            b. $\sqrt[3]{48a^4b^5c^6} = \sqrt[3]{8 \cdot 6 \cdot a^3 \cdot a \cdot b^3 \cdot b^2 \cdot c^6} = \sqrt[3]{8a^3b^3c^6} \cdot \sqrt[3]{6ab^2} = 2abc^2 \sqrt[3]{6ab^2}$

            c. $\dfrac{\sqrt[4]{160x^7}}{\sqrt[4]{10x^3}} = \sqrt[4]{\dfrac{160x^7}{10x^3}} = \sqrt[4]{16x^4} = 2x$

EXAMPLE 3   *Objective 4: Use rational exponents to simplify radical expressions.*

Use rational exponents to simplify $\sqrt[4]{7} \cdot \sqrt{200}$.

Solution: $\sqrt[4]{7} = 7^{1/4}$

$\sqrt{200} = 200^{1/2} = 200^{2/4} = \sqrt[4]{200^2} = \sqrt[4]{40{,}000}$

$\sqrt[4]{7} \cdot \sqrt{200} = \sqrt[4]{7} \cdot \sqrt[4]{40{,}000} = \sqrt[4]{280{,}000} = \sqrt[4]{28} \cdot \sqrt[4]{10{,}000} = 10\sqrt[4]{28}$

EXERCISES

Exercises 1–6. Perform the indicated operation. Simplify if possible. See Example 1.

1. $\sqrt[3]{9} \cdot \sqrt[3]{3}$

2. $\sqrt[4]{x^2 y} \cdot \sqrt[4]{xy^3}$

3. $\sqrt[5]{5a^2b^3} \cdot \sqrt[5]{2ab}$

4. $\dfrac{\sqrt[3]{90}}{\sqrt[3]{10}}$

5. $\dfrac{\sqrt{60x^4}}{\sqrt{15x^3}}$

6. $\dfrac{\sqrt[3]{p^6 q^5}}{\sqrt[3]{p^5 q^2}}$

Exercises 7–12. Simplify. See Example 2.

7. $\sqrt[3]{81}$

8. $\sqrt{160}$

9. $\sqrt[5]{40x^7 y^4 z^5}$

10. $\sqrt[3]{-125a^{15}b^{10}c^8}$

11. $\sqrt{\dfrac{17x^5 y}{68x^2 y^3}}$

12. $\sqrt[3]{\dfrac{a^6 b^{11}}{64}}$

Exercises 12–13. Use rational exponents to write each radical with the same index. Then multiply. See Example 3.

13. $\sqrt{3} \cdot \sqrt[3]{4}$

14. $\sqrt[5]{2x} \cdot \sqrt[3]{6}$

15. The formula for the surface area A of a cone with height h and radius r is given by $A = \pi r \sqrt{r^2 + h^2}$. Find the surface area of a cone whose height is 12 inches and whose radius is 5 inches.

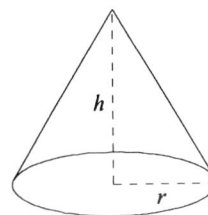

16. The demand equation for news magazines sold through newsstand purchases is $F(x) = 5\sqrt{6 - 0.8x}$, where x is the newsstand price per magazine and $F(x)$ is the quantity demanded per year in hundreds of thousands. Approximate the annual demand if the magazine is priced at \$2.50 per copy at the newsstand.

7.4   Adding, Subtracting, and Multiplying Radical Expressions

EXAMPLE 1   *Objective 1: Add or subtract radical expressions.*

Perform the indicated operation.

a. $8\sqrt[3]{4}+15\sqrt[3]{32}$ b. $9x\sqrt[4]{x^2y^3}-2\sqrt[4]{16x^6y^3}$

Solution: a. $8\sqrt[3]{4}+15\sqrt[3]{32}=8\cdot\sqrt[3]{4}+15\cdot\sqrt[3]{8}\cdot\sqrt[3]{4}=8\sqrt[3]{4}+30\sqrt[3]{4}=38\sqrt[3]{4}$

b. $9x\sqrt[4]{x^2y^3}-2\sqrt[4]{16x^6y^3}=9\cdot x\cdot\sqrt[4]{x^2y^3}-2\cdot\sqrt[4]{16x^4}\cdot\sqrt[4]{x^2y^3}$

$$=9x\sqrt[4]{x^2y^3}-4x\sqrt[4]{x^2y^3}$$

$$=5x\sqrt[4]{x^2y^3}$$

EXAMPLE 2   *Objective 2: Multiply radical expressions.*

Multiply.

a. $\sqrt{5}(6+\sqrt{5})$ b. $(\sqrt{2q}+1)(3\sqrt{q}-4)$

Solution: a. $\sqrt{5}(6+\sqrt{5})=\sqrt{5}(6)+\sqrt{5}\left(\sqrt{5}\right)=6\sqrt{5}+\sqrt{25}=6\sqrt{5}+5$

b. $(\sqrt{2q}+1)(3\sqrt{q}-4)=\sqrt{2q}\cdot3\sqrt{q}-4\cdot\sqrt{2q}+1\cdot3\sqrt{q}-4\cdot1$

$$=3q\sqrt{2}-4\sqrt{2q}+3\sqrt{q}-4$$

EXERCISES

Exercises 1–8. Perform the indicated operation. Simplify if possible. See Example 1.

1. $\sqrt{50}-\sqrt{72}$ 2. $\sqrt{20}+\sqrt{45}$ 3. $7a\sqrt[4]{ab}-\sqrt[4]{16a^5b}$ 4. $10\sqrt[3]{2x^4y^2}+3x\sqrt[3]{2xy^2}$

5. $\dfrac{6\sqrt{5}}{5}-\dfrac{\sqrt{5}}{5}$ 6. $\dfrac{2\sqrt{3}}{3}+\dfrac{5\sqrt{3}}{4}$ 7. $\dfrac{\sqrt[3]{32x^2}}{2}+\dfrac{\sqrt[3]{4x^2}}{6}$ 8. $\sqrt{\dfrac{x^3}{25}}-\dfrac{\sqrt{x^5}}{10}$

Exercises 9–16. Multiply and then simplify, if possible. See Example 2.

9. $\sqrt{3}\left(\sqrt{6}+\sqrt{10}\right)$ 10. $\left(\sqrt{5}-\sqrt{8}\right)^2$ 11. $\sqrt{10x}\left(\sqrt{2}+\sqrt{x}\right)$ 12. $\left(2\sqrt{x}+\sqrt{3}\right)\left(2\sqrt{x}-\sqrt{3}\right)$

13. $\left(2\sqrt{2y}+\sqrt{3y}\right)\left(2-\sqrt{y}\right)$ 14. $\left(a\sqrt{2}+\sqrt{6}\right)\left(a-\sqrt{3}\right)$

15. $\sqrt[3]{3x}\left(\sqrt[3]{9x^2}+\sqrt[3]{2}\right)$

16. $\left(\sqrt[3]{x}+1\right)\left(\sqrt[3]{x^2}-1\right)$

17. Find the perimeter of the triangle. (All measurements are given in centimeters.)

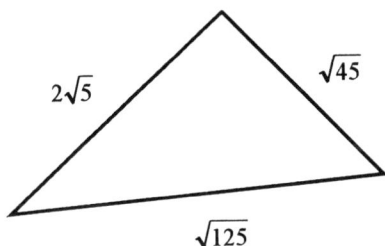

18. Find the area of the rectangle.

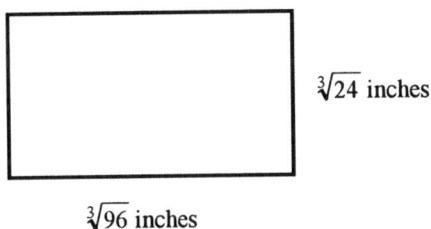

7.5   Rationalizing Numerators and Denominators of Radical Expressions

EXAMPLE 1   *Objective 1: Rationalize denominators.*

Rationalize the denominator of each expression.

a. $\dfrac{20a}{\sqrt{28b}}$ **b.** $\dfrac{\sqrt{2q}}{\sqrt{5q}+\sqrt{10r}}$

Solution: **a.**   First, simplify $\dfrac{20a}{\sqrt{28b}}$; then rationalize the denominator.

$$\frac{20a}{\sqrt{28b}}=\frac{20a}{\sqrt{4\cdot 7\cdot b}}=\frac{20a}{2\sqrt{7b}}=\frac{10a}{\sqrt{7b}}=\frac{10a\cdot\sqrt{7b}}{\sqrt{7b}\cdot\sqrt{7b}}=\frac{10a\sqrt{7b}}{7b}$$

b.   Multiply the numerator and denominator by the conjugate of $\sqrt{5q}+\sqrt{10r}$.

$$\frac{\sqrt{2q}}{\sqrt{5q}+\sqrt{10r}}=\frac{\sqrt{2q}\cdot\left(\sqrt{5q}-\sqrt{10r}\right)}{\left(\sqrt{5q}+\sqrt{10r}\right)\cdot\left(\sqrt{5q}-\sqrt{10r}\right)}$$

$$=\frac{\sqrt{2q}\sqrt{5q}-\sqrt{2q}\sqrt{10r}}{\left(\sqrt{5q}\right)^2-\left(\sqrt{10r}\right)^2}$$

$$= \frac{\sqrt{10q^2} - \sqrt{20qr}}{5q - 10r}$$

$$= \frac{q\sqrt{10} - 2\sqrt{5qr}}{5q - 10r}$$

EXAMPLE 2   *Objective 2: Rationalize numerators.*

Rationalize the numerator of each expression.

a. $\dfrac{\sqrt{12a}}{2}$ **b.** $\dfrac{3\sqrt{y} - 5\sqrt{2y}}{2y}$

Solution: a. First, simplify $\dfrac{\sqrt{12a}}{2}$; then rationalize the numerator.

$$\frac{\sqrt{12a}}{2} = \frac{\sqrt{4 \cdot 3 \cdot a}}{2} = \frac{2\sqrt{3a}}{2} = \frac{\sqrt{3a}}{1} = \frac{\sqrt{3a} \cdot \sqrt{3a}}{1 \cdot \sqrt{3a}} = \frac{3a}{\sqrt{3a}}$$

b. Multiply the numerator and denominator by the conjugate of $3\sqrt{y} - 5\sqrt{2y}$.

$$\frac{3\sqrt{y} - 5\sqrt{2y}}{2y} = \frac{\left(3\sqrt{y} - 5\sqrt{2y}\right)\left(3\sqrt{y} + 5\sqrt{2y}\right)}{2y\left(3\sqrt{y} + 5\sqrt{2y}\right)}$$

$$= \frac{\left(3\sqrt{y}\right)^2 - \left(5\sqrt{2y}\right)^2}{2y\left(3\sqrt{y} + 5\sqrt{2y}\right)}$$

$$= \frac{9y - 50y}{2y\left(3\sqrt{y} + 5\sqrt{2y}\right)}$$

$$= \frac{-41y}{2y\left(3\sqrt{y} + 5\sqrt{2y}\right)}$$

EXERCISES

Exercises 1–8. Simplify by rationalizing the denominator. See Example 1.

1. $\dfrac{\sqrt[3]{4}}{\sqrt[3]{9}}$ **2.** $\dfrac{7}{\sqrt{35b}}$ **3.** $\dfrac{12}{1 - \sqrt{3}}$ **4.** $\dfrac{36}{4 + \sqrt{22}}$

5. $\dfrac{6y}{\sqrt{y} + 4}$ **6.** $\dfrac{\sqrt{3} - y}{\sqrt{3} + y}$ **7.** $\dfrac{\sqrt{2z} + 3}{\sqrt{2z} - 15}$ **8.** $\dfrac{\sqrt{3x}}{\sqrt{x} - \sqrt{3x}}$

Exercises 9–16. Rationalize each numerator. See Example 2.

9. $\sqrt{\dfrac{20}{3}}$

10. $\dfrac{\sqrt[3]{4}}{\sqrt[3]{9}}$

11. $\dfrac{\sqrt{5a}}{15}$

12. $\dfrac{\sqrt[3]{18z^2}}{5}$

13. $\dfrac{\sqrt{3x}}{x-\sqrt{3x}}$

14. $\dfrac{\sqrt{8x^3}}{5x+\sqrt{2x}}$

15. $\dfrac{\sqrt{3}-y}{\sqrt{3}+y}$

16. $\dfrac{\sqrt{2z}+3}{\sqrt{2z}-15}$

C *Exercises 17–18. Use a grapher to verify the solution of each given exercise.*

17. Exercise 7

18. Exercise 15

7.6 Radical Equations and Problem Solving

EXAMPLE 1   *Objective 1: Solve equations containing radical expressions.*

Solve $\sqrt{35-2x}+x=2x$ for x.

Solution: First, isolate the radical on one side of the equation. To do this, subtract x from both sides.

$$\sqrt{35-2x}+x=2x$$
$$\sqrt{35-2x}=x$$

Next, use the power rule to eliminate the radical.

$$\left(\sqrt{35-2x}\right)^2=(x)^2$$
$$35-2x=x^2$$
$$x^2+2x-35=0$$
$$(x+7)(x-5)=0$$
$$x+7=0 \quad \text{or} \quad x-5=0$$
$$x=-7 \qquad x=5$$

A check of the possible solutions shows that -7 is extraneous. Thus, the solution set is $\{5\}$.

EXAMPLE 2   *Objective 1: Solve equations containing radical expressions.*

Solve $\sqrt{y+3}+\sqrt{y}=5$ for y.

Solution: First, isolate the radical by subtracting $\sqrt{y}$ from both sides.

$$\sqrt{y+3} + \sqrt{y} = 5$$
$$\sqrt{y+3} = 5 - \sqrt{y}$$

Use the power rule to begin eliminating radicals. Square both sides.

$$\left(\sqrt{y+3}\right)^2 = \left(5 - \sqrt{y}\right)^2$$
$$y + 3 = 25 - 10\sqrt{y} + y$$
$$10\sqrt{y} = 22$$

Use the power rule again to eliminate the remaining radical.

$$\left(10\sqrt{y}\right)^2 = (22)^2$$
$$100y = 484$$
$$y = 4.84$$

The proposed solution, 4.84, does check in the original equation, so the solution set is {4.84}.

EXAMPLE 3   *Objective 2: Use the Pythagorean Theorem to model problems.*

Find the length of the length of the hypotenuse of the following right triangle.

14 yards

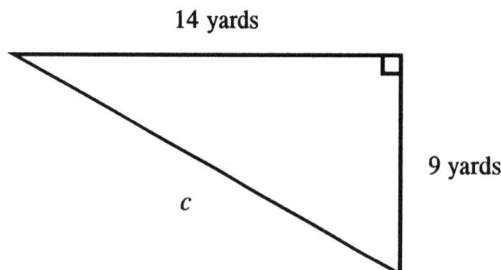

9 yards

c

Solution: Let $a = 14$, $b = 9$, and solve for c.

$$a^2 + b^2 = c^2$$
$$14^2 + 9^2 = c^2$$
$$196 + 81 = c^2$$
$$277 = c^2$$
$$c = \sqrt{277} \quad \text{or} \quad c = -\sqrt{277}$$

Because we are solving for a length, we will list the positive solution only. The hypotenuse of the triangle is $\sqrt{277} \approx 16.64$ yards long.

EXERCISES

Exercises 1–2. Solve each equation.

1. $\sqrt{3x - 8} = 8$

2. $\sqrt{5y - 16} - 4 = 3$

Exercises 3–4. Solve each equation. See Example 1.

3. $t + \sqrt{t+2} = 0$

4. $5x = 3x + \sqrt{x+33}$

Exercises 5–8. Solve each equation.

5. $\sqrt[3]{2x+6} - 1 = 7$

6. $\sqrt[3]{5b-3} + 6 = 9$

7. $2z + 3 = \sqrt{5z+9}$

8. $4x - 2 = 3x + \sqrt{1-8x}$

Exercises 9–10. Solve each equation. See Example 2.

9. $\sqrt{x-4} - \sqrt{x+4} = -2$

10. $\sqrt{3y-5} = \sqrt{5y-6} - 1$

Exercises 11–12. Find the length of the unknown side in each triangle.

11.

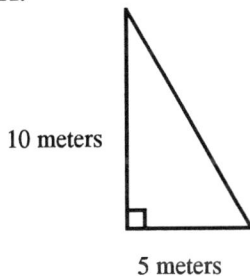

10 meters

5 meters

12.

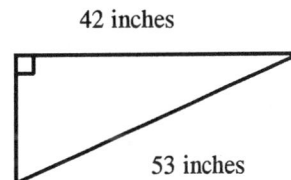

42 inches

53 inches

13. The demand equation for a certain news magazine sold through newsstand purchases is $F(x) = 5\sqrt{6 - 0.8x}$, where x is the newsstand price per magazine and $F(x)$ is the quantity demanded per year in hundreds of thousands. Find the greatest price that can be charged for the magazine so that at least 8 hundred thousand magazines will be sold annually.

14. A visitor to a zoo has accidentally dropped his keys into the bear exhibit. The zookeeper has volunteered to retrieve the keys. To do so, the zookeeper must use a board to form a ramp over the moat surrounding the bear exhibit. What is the minimum length the board must be to reach over the moat? See figure.

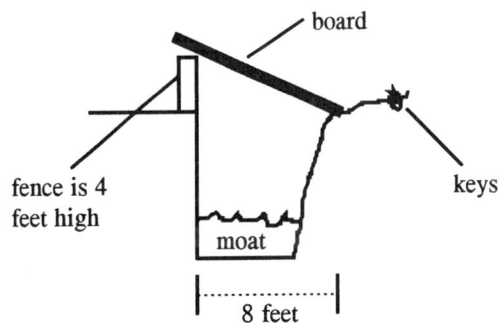

board

fence is 4 feet high

keys

moat

8 feet

7.7   Complex Numbers

EXAMPLE 1   *Objective 2: Add or subtract complex numbers.*

Add or subtract the complex numbers. Write the sum or difference in the form $a + bi$.
a. $(6 + 5i) + (-13 + 3i)$ **b.** $(9 + 4i) - (2 - 8i)$

Solution: **a.** $(6 + 5i) + (-13 + 3i) = (6 - 13) + (5 + 3)i = -7 + 8i$
b. $(9 + 4i) - (2 - 8i) = 9 + 4i - 2 + 8i = (9 - 2) + (4 + 8)i = 7 + 12i$

EXAMPLE 2   *Objective 3: Multiply complex numbers.*

Multiply the complex numbers. Write the product in the form $a + bi$.
a. $(1 + 9i)(3 - 4i)$ **b.** $(5 - 2i)(5 + 2i)$

Solution: **a.** $(1 + 9i)(3 - 4i) = 1(3) - 1(4i) + 9i(3) - 9i(4i)$
$$= 3 - 4i + 27i - 36i^2$$
$$= 3 + 23i - 36(-1)$$
$$= 3 + 23i + 36$$
$$= 39 + 23i$$
b. $(5 - 2i)(5 + 2i) = 5^2 - (2i)^2 = 25 - 4i^2 = 25 - 4(-1) = 25 + 4 = 29$

EXAMPLE 3   *Objective 4: Divide complex numbers.*

Find each quotient. Write in the form $a + bi$.

a. $\dfrac{14}{5i}$ **b.** $\dfrac{10 - 3i}{5 + 4i}$

Solution: **a.**   Multiply the numerator and the denominator by the complex conjugate of $5i$ to eliminate the imaginary number in the denominator.
$$\frac{14}{5i} = \frac{14(-5i)}{5i(-5i)} = \frac{-70i}{-25i^2} = \frac{-70i}{-25(-1)} = \frac{-70i}{25} = -\frac{14i}{5}$$
In $a + bi$ form, this is $0 - \dfrac{14i}{5}$.

b.   Multiply the numerator and the denominator by the complex conjugate of $5 + 4i$.
$$\frac{10 - 3i}{5 + 4i} = \frac{(10 - 3i)(5 - 4i)}{(5 + 4i)(5 - 4i)}$$
$$= \frac{50 - 40i - 15i + 12i^2}{5^2 - (4i)^2}$$
$$= \frac{50 - 55i + 12(-1)}{25 - 16i^2}$$
$$= \frac{50 - 55i - 12}{25 - 16(-1)}$$
$$= \frac{38 - 55i}{25 + 16}$$
$$= \frac{38 - 55i}{41} = \frac{38}{41} - \frac{55}{41}i$$

EXERCISES

Exercises 1–4. Add or subtract. Write the result in the form $a + bi$. See Example 1.

1. $(1+i)-(3-i)$ **2.** $(2-i)+(-9-3i)$ **3.** $(18+7i)+(-15+2i)$ **4.** $(2-4i)-(5-8i)$

Exercises 5–12. Multiply or divide. Write the result in the form $a + bi$. See Examples 2 and 3.

5. $(-3i)(20i)$ **6.** $(-11i)(-12i)$ **7.** $2i(7+15i)$ **8.** $(14-2i)(5+3i)$

9. $(2-7i)(2+7i)$ **10.** $\dfrac{10}{1+3i}$ **11.** $\dfrac{1+9i}{3-2i}$ **12.** $\dfrac{12-1i}{1-5i}$

Exercises 13–16. Multiply or divide.

13. $\sqrt{-6}\cdot\sqrt{-8}$ **14.** $\sqrt{-10}\cdot\sqrt{-12}$ **15.** $\sqrt{-7}\cdot\sqrt{8}$ **16.** $\dfrac{\sqrt{-90}}{\sqrt{-6}}$

Exercises 17–20. Find each power of i.

17. i^{13} **18.** i^{42} **19.** i^{28} **20.** i^{37}

Chapter 7   Hints and Warnings

- To **simplify a radical expression with index n,** be sure to remove all factors that contain perfect nth powers.
- To **add or subtract radical expressions,** remember that only terms with like radicals can be combined.
- To **multiply radical expressions,** remember that many of the same methods as those used when multiplying polynomials may be used. *For example, the FOIL method may be helpful.*
- To **solve a radical equation:**
    - Step 1: Write the equation so that one radical is by itself on one side of the equation.
    - Step 2: Raise each side of the equation to a power equal to the index of the radical.
    - Step 3: Simplify each side of the equation.
    - Step 4: If the equation still contains a radical, repeat Steps 1 through 3.
    - Step 5: Solve the equation.
    - Step 6: Check the proposed solution(s) in the original equation. Be sure to eliminate any extraneous solutions from the solution set.
- To **add or subtract complex numbers,** add or subtract their real parts and then add or subtract their imaginary parts.
- To **multiply complex numbers,** multiply as though they are binomials. *The FOIL method will be useful.*
- To **divide complex numbers,** multiply the numerator and denominator by the complex conjugate of the denominator.

CHAPTER 7   Practice Test

Raise to the power or take the root. Assume that all variables represent positive numbers. Write using only positive exponents.

1. $\sqrt{343}$

2. $-\sqrt[3]{x^{33}}$

3. $\left(\dfrac{1}{256}\right)^{1/4}$

4. $\left(\dfrac{1}{256}\right)^{-1/4}$

5. $\left(\dfrac{81x^4}{16}\right)^{3/4}$

6. $\sqrt[3]{-q^{24}r^{39}}$

7. $\left(\dfrac{36z^{3/4}}{x^{1/3}y^{-1/6}}\right)^{1/2}$

8. $y^{3/2}\left(y^{8/3}-y^{-1/2}\right)$

Take the root. Use absolute value bars when necessary.

9. $\sqrt[5]{(13ab)^5}$

10. $\sqrt{(7x)^2}$

Rationalize the denominator.

11. $\sqrt{\dfrac{25}{3x}}$

12. $\dfrac{5+2\sqrt{y}}{6-\sqrt{y}}$

13. $\dfrac{\sqrt[3]{5x^2y^2}}{\sqrt[3]{xyz}}$

Perform the indicated operations. Assume that all variables represent positive numbers.

14. $\sqrt[3]{81y^5}-5y\sqrt[3]{24y^2}$

15. $\sqrt{2}\left(\sqrt{50}-\sqrt{13}\right)$

16. $\left(\sqrt{2a}-4\right)^2$

17. $\left(\sqrt{3}+5\right)\left(\sqrt{7}-1\right)$

18. $\left(\sqrt{6}-2\right)\left(\sqrt{6}+2\right)$

Use a calculator to approximate each to 3 decimal places.

19. $\sqrt{362}$

20. $283^{-3/4}$

Solve each equation.

21. $x = \sqrt{2x+3}$

22. $\sqrt{4y^2 - 20y + 50} + 5 = 0$

23. $\sqrt{3x-4} = \sqrt{x+6}$

Perform the indicated operation and simplify. Write the result in the form $a + bi$.

24. $\sqrt{-9}$

25. $-\sqrt{-18}$

26. $(10 - 3i) - (5 - 6i)$

27. $(3 + 3i)(3 - 3i)$

28. $(5 - 7i)^2$

29. $\dfrac{8+i}{2-2i}$

30. Find x.

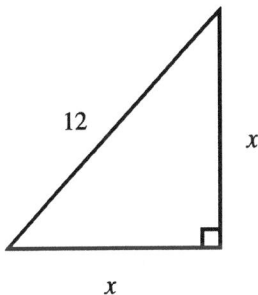

31. Identify the domain of $f(x)$. Then complete the table below and graph $f(x)$.

$$f(x) = \sqrt{2x-4}$$

x	2	2.5	4	6
$f(x)$				

32. The demand function for a new product is $f(x) = 2\sqrt{50-x}$, where x is the price of the product and $f(x)$ is the number of units (in thousands) sold at each price.
   (a) How many units will be sold when the price is $25?
   (b) At what price will 12,000 units be sold?

CHAPTER 8

Quadratic Equations and Functions

8.1 Solving Quadratic Equations by Completing the Square

EXAMPLE 1 *Objective 1: Use the square root property to solve quadratic equations.*

Use the square root property to solve $(3x+8)^2 = 2$.

Solution: By the square root property, we have

$$(3x+8)^2 = 2$$

$$3x+8 = \pm\sqrt{2} \qquad \text{Apply the square root property.}$$

$$3x = -8 \pm\sqrt{2} \qquad \text{Subtract 8 from both sides.}$$

$$x = \frac{-8 \pm\sqrt{2}}{3} \qquad \text{Divide both sides by 3.}$$

The solution set is $\left\{ \dfrac{-8+\sqrt{2}}{3}, \dfrac{-8-\sqrt{2}}{3} \right\}$.

EXAMPLE 2 *Objective 3: Solve quadratic equations by completing the square.*

Solve $y^2 + 5y + 3 = 0$ for y by completing the square.

Solution: First, subtract 3 from both sides of the equation so that the left side has no constant term.

$$y^2 + 5y + 3 = 0$$

$$y^2 + 5y = -3$$

Now, find the constant term that makes the left side a perfect square trinomial by squaring half the coefficient of the y-term. Add this constant to *both* sides of the equation.

$$\frac{1}{2}(5) = \frac{5}{2} \quad \text{and} \quad \left(\frac{5}{2}\right)^2 = \frac{25}{4}$$

$$y^2 + 5y + \frac{25}{4} = -3 + \frac{25}{4} \qquad \text{Add } \frac{25}{4} \text{ to both sides of the equation.}$$

$$\left(y + \frac{5}{2}\right)^2 = \frac{13}{4} \qquad \text{Factor the perfect square trinomial and simplify the right side.}$$

$$y + \frac{5}{2} = \pm\sqrt{\frac{13}{4}} \qquad \text{Apply the square root property.}$$

$$y = -\frac{5}{2} \pm \frac{\sqrt{13}}{2} \qquad \text{Subtract } \frac{5}{2} \text{ from both sides and simplify } \sqrt{\frac{13}{4}}.$$

$$y = \frac{-5 \pm \sqrt{13}}{2} \qquad \text{Simplify.}$$

The solution set is $\left\{ \dfrac{-5+\sqrt{13}}{2}, \dfrac{-5-\sqrt{13}}{2} \right\}$.

EXAMPLE 3   *Objective 3: Solve quadratic equations by completing the square.*

Solve $3x^2 + 24x + 72 = 0$.

Solution: Our procedure for finding the constant term to complete the square works only if the coefficient of the squared variable term is 1. Therefore, to solve this equation, the first step is to divide both sides by 3, the coefficient of x^2.

$$3x^2 + 24x + 72 = 0$$

$$x^2 + 8x + 24 = 0 \qquad \text{Divide both sides by 3.}$$

$$x^2 + 8x = -24 \qquad \text{Subtract 24 from both sides.}$$

Next, find the square of half of 8.

$$\frac{1}{2}(8) = 4 \quad \text{and} \quad 4^2 = 16$$

$$x^2 + 8x + 16 = -24 + 16 \qquad \text{Add 16 to both sides of the equation.}$$

$$(x - 4)^2 = -8 \qquad \text{Factor the perfect square and simplify the right side.}$$

$$x - 4 = \pm\sqrt{-8} \qquad \text{Apply the square root property.}$$

$$x - 4 = \pm 2i\sqrt{2} \qquad \text{Simplify the radical.}$$

$$x = 4 \pm 2i\sqrt{2} \qquad \text{Add 4 to both sides.}$$

The solution set is $\left\{ 4 + 2i\sqrt{2},\ 4 - 2i\sqrt{2} \right\}$.

EXERCISES

Exercises 1–8. Use the square root property to solve each equation. See Example 1.

1. $w^2 = 24$

2. $x^2 = 40$

3. $q^2 - \dfrac{25}{4} = 0$

4. $a^2 + 14 = 50$

5. $3y^2 + 48 = 0$

6. $(z - 4)^2 = 18$

7. $(2r - 1)^2 + 12 = 0$

8. $(3t + 2)^2 = -25$

Exercises 9–10. Add the proper constant to each binomial so that the resulting trinomial is a perfect square trinomial. Then factor the trinomial.

9. $x^2 - 13x$

10. $b^2 + 22b$

Exercises 11–18. Solve each equation by completing the square. See Examples 2 and 3.

11. $x^2 + 6x - 11 = 0$

12. $y^2 - 3y + 7 = 0$

13. $z^2 + 2z = -6$ **14.** $t^2 - 4t - 15 = 0$

15. $2x^2 + 10x - 3 = 0$ **16.** $2a^2 - 6a - 20 = 0$

17. $3t^2 - 9t + 36 = 0$ **18.** $5y^2 + 10y + 100 = 0$

Exercises 19–20. Use the formula $A = P(1+r)^t$ *to solve each exercise.*

19. Find the rate r to make \$5000 grow to \$5724.50 in 2 years.

20. Find the rate r to make \$4000 grow to \$5017.60 in 2 years.

Exercises 21–22. Neglecting air resistance, the distance traveled by a freely falling object is given by the function $s(t) = 16t^2$, *where t is time in seconds. Use this formula to solve each exercise.*

21. The NationsBank Tower in Atlanta, Georgia, is 1023 feet tall. How long would it take an object to fall from the top of the building and hit the ground? (data from *The World Almanac*)

22. The Terminal Tower in Cleveland, Ohio, is 708 feet tall. How long would it take an object to fall from the top of the building? (data from *The World Almanac*)

23. The area of the square dance floor in a large hotel is 1089 square feet. Find the dimensions of the room.

24. The area of a circle is 324π square centimeters. Find the radius of the circle.

25. A sporting goods store has found that the demand equation for a certain type of hockey stick is $p = -x^2 + 60$, where p represents the unit price and x represents the quantity demanded in thousands. Find the demand for the hockey stick if the price is \$22 per stick.

C *Exercises 26–27. Use a grapher to solve each equation graphically. Round all solutions to two decimal places.*

26. $x^2 + 3x = 2x + 15$ **27.** $-3x^2 + 8x = -21$

8.2   Solving Quadratic Equations by the Quadratic Formula

EXAMPLE 1   *Objective 1: Solve quadratic equations using the quadratic formula.*

Solve $2x^2 + 5x = 13$.

Solution: First, write the equation in standard form by subtracting 13 from both sides.

$$2x^2 + 5x = 13$$
$$2x^2 + 5x - 13 = 0$$

Now $a = 2$, $b = 5$, and $c = -13$. Substitute these values into the quadratic formula.

$$x = \frac{-b \pm \sqrt{b^2 - 4ac}}{2a}$$
$$= \frac{-5 \pm \sqrt{5^2 - 4(2)(-13)}}{2(2)}$$
$$= \frac{-5 \pm \sqrt{25 + 104}}{4}$$
$$= \frac{-5 \pm \sqrt{129}}{4}$$

The solution set is $\left\{ \dfrac{-5 + \sqrt{129}}{4}, \dfrac{-5 - \sqrt{129}}{4} \right\}$.

EXAMPLE 2   *Objective 2: Determine the number and type of solutions of a quadratic equation using the discriminant.*

Use the discriminant to determine the number and type of solutions of each quadratic equation.
a. $x^2 + 3x + 5 = 0$ **b.** $y^2 + 6x - 7 = 0$ **c.** $4x^2 - 20x + 25 = 0$

Solution: **a.**   For $x^2 + 3x + 5 = 0$, $a = 1$, $b = 3$, and $c = 5$. Thus, $b^2 - 4ac = 3^2 - 4(1)(5) = 9 - 20 = -11$. Because $b^2 - 4ac$ is negative, the quadratic equation has two complex solutions but not real solutions.

b.   For $y^2 + 6x - 7 = 0$, $a = 1$, $b = 6$, and $c = -7$. Thus, $b^2 - 4ac = 6^2 - 4(1)(-7) = 36 + 28 = 64$. Because $b^2 - 4ac$ is positive, the quadratic equation has two real solutions.

c.   For $4x^2 - 20x + 25 = 0$, $a = 4$, $b = -20$, and $c = 25$. Thus, $b^2 - 4ac = (-20)^2 - 4(4)(25) = 400 - 400 = 0$. Because $b^2 - 4ac$ is zero, the quadratic equation has only one real solution.

EXERCISES

Exercises 1–12. Use the quadratic formula to solve each quadratic equation. See Example 1.

1. $x^2 + 2x - 3 = 0$ **2.** $y^2 - 8y - 9 = 0$

3. $3z^2 + 6z + 2 = 0$

4. $2a^2 + 5a + 2 = 0$

5. $5a^2 - 3a - 10 = 0$

6. $4q^2 + q - 3 = 0$

7. $y^2 + y + 1 = 0$

8. $t^2 + 7t + 8 = 0$

9. $2a^2 - 6a - 20 = 0$

10. $5y^2 + 10y + 100 = 0$

11. $3t^2 - 9t + 36 = 0$

12. $t^2 - 4t - 15 = 0$

Exercises 13–16. Use the discriminant to determine the number and types of solutions of each equation. See Example 2.

13. $2x^2 - 9x - 17 = 0$

14. $4a^2 + 20a + 3 = 0$

15. $3y^2 + y + 4 = 0$

16. $d^2 - 6d + 9 = 0$

17. A ball is thrown downward from the top of the 546-foot tall TCBY Towers in Little Rock, Arkansas, with an initial velocity of -15 feet per second. The height of the ball (in feet) after t seconds is given by the function $h(t) = -16t^2 - 15t + 546$. How long after the ball is thrown will it strike the ground?

18. The graph shows the amount given to charity (in billions of inflation-adjusted dollars) by corporations over the period 1984 through 1992. (data from American Association of Fund-Raising Counsel, Inc.)

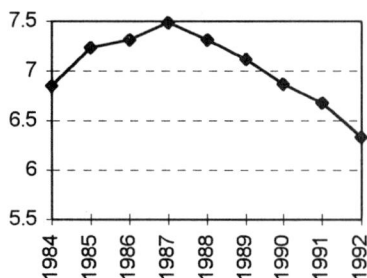

(a) According to the graph, when was charitable giving by corporations the highest?

(b) Use the graph to estimate the level of charitable giving by corporations in 1990.

(c) This data can be modeled by the quadratic function
$f(x) = -0.045x^2 + 0.632x + 5.12$, where $f(x)$ is the level of charitable giving in billions of dollars and x is the number of years since 1980. Use the model to find the level of giving in 1987. Does this agree with the graph?

(d) Use the model and the quadratic formula to find the year(s) in which giving was $7.296 billion. Does this agree with the graph?

8.3   Solving Equations Using Quadratic Methods

EXAMPLE 1   *Objective 1: Solve various equations that are quadratic in form.*

Solve $x^4 + 6x^2 - 27 = 0$.

Solution: First, factor the trinomial.

$$x^4 + 6x^2 - 27 = 0$$
$$(x^2 - 3)(x^2 + 9) = 0 \qquad\qquad \text{Factor.}$$
$$x^2 - 3 = 0 \quad \text{or} \quad x^2 + 9 = 0 \qquad \text{Set each factor equal to 0.}$$
$$x^2 = 3 \quad\;\; \text{or} \quad x^2 = -9$$
$$x = \pm\sqrt{3} \quad \text{or} \quad x = \pm 3i$$

The solution set is $\left\{ \sqrt{3}, -\sqrt{3}, 3i, -3i \right\}$.

EXAMPLE 2   *Objective 1: Solve various equations that are quadratic in form.*

Solve $(x+2)^4 + 6(x+2)^2 - 27 = 0$.

Solution: Notice that the quantity $(x+2)$ is repeated in this equation. Let $y = (x+2)$ and substitute in the original equation.

$$(x+2)^4 + 6(x+2)^2 - 27 = 0$$
$$y^4 + 6y^2 - 27 = 0 \qquad \text{Let } y = x+2.$$

Notice that this equation in y is the same as the equation we solved in Example 1. Thus,

$$y = \pm\sqrt{3} \quad \text{and} \quad x+2 = \pm\sqrt{3}$$
$$x = -2 \pm \sqrt{3}$$

$$y = \pm 3i \quad \text{and} \quad x+2 = \pm 3i$$
$$x = -2 \pm 3i$$

Thus, the solution set is $\left\{ -2+\sqrt{3}, -2-\sqrt{3}, -2+3i, -2-3i \right\}$.

EXERCISES

Exercises 1–8. Solve.

1. $\dfrac{1}{x} - \dfrac{4}{x-1} = 2$

2. $\dfrac{5}{x+1} + \dfrac{3}{x-1} = -1$

3. $y^4 + 6y^2 + 5 = 0$

4. $z^4 + 7y^2 + 12 = 0$

5. $(2x-4)^2 - 3(2x-4) + 2 = 0$

6. $(q+3)^2 - 5(q+3) + 8 = 0$

7. $x^{2/3} + x^{1/3} - 6 = 0$

8. $t^{2/3} - 2t^{1/3} - 8 = 0$

9. Two printing presses can print a 3000-item print run in 5 hours working together. When working alone, printing press #1 can finish a 3000-item print run in 1.5 hours less than printing press #2 can. Find how long to the nearest hundredth-hour it takes each printing press to complete a 3000-item print run working alone.

10. The first and second floors of an office building share photocopier supplies. Together, the two floors can use 100 reams of photocopy paper in 50 days. By itself, the second floor uses 100 reams of paper in 20 days more than the time it takes the first floor to use 100 reams of paper. How many days to the nearest whole day would it take for the first floor to use up 100 reams of photocopy paper by itself?

C *Exercises 11–12. Use a grapher to solve each exercise graphically. Round all solutions to two decimal places. Compare the solutions with the solution you obtained algebraically. Explain any differences.*

11. Exercise 4

12. Exercise 5

8.4 Nonlinear Inequalities in One Variable

EXAMPLE 1 *Objective 1: Polynomial inequalities of degree 2 or greater.*

Solve $x^2 - 3x - 28 \le 0$.

Solution: First, solve the corresponding equation $x^2 - 3x - 28 = 0$.

$$x^2 - 3x - 28 = 0$$
$$(x-7)(x+4) = 0$$
$$x - 7 = 0 \quad \text{or} \quad x + 4 = 0$$
$$x = 7 \qquad\qquad x = -4$$

These solutions separate the number line into three intervals. Next, check test points in each interval.

		Test Point	$(x-7)(x+4) \le 0$	
Region A:	$(-\infty, -4]$	-5	$(-12)(-1) \le 0$	False
Region B:	$[-4, 7]$	0	$(-7)(4) \le 0$	True
Region C:	$[7, \infty)$	10	$(3)(14) \le 0$	False

The solution set is $[-4, 7]$.

EXAMPLE 2   *Objective 2: Solve inequalities containing rational expressions with variables in the denominator.*

Solve $\dfrac{3x - 12}{x + 4} \geq 0$.

Solution: First, find the values of x that make the denominator equal to 0.

$$x + 4 = 0$$
$$x = -4$$

Next, solve $\dfrac{3x - 12}{x + 4} = 0$.

$$\dfrac{3x - 12}{x + 4} = 0 \qquad \text{Multiply both sides by the LCD, } x + 4.$$
$$3x - 12 = 0$$
$$3x = 12$$
$$x = 4$$

Use these two solutions to divide a number line into three intervals: $(-\infty, -4)$, $(-4, 4]$, and $[4, \infty)$, and choose test points. Test point values from both the first and third intervals satisfy the original inequality. The solution set is $(-\infty, -4) \cup [4, \infty)$.

EXERCISES

Exercises 1–6. Solve each quadratic inequality. Write the solution set in interval notation. See Example 1

1. $(2x - 3)(x + 1) > 0$

2. $(x - 7)(x - 9) > 0$

3. $(x + 9)(3x - 5) < 0$

4. $(2x + 10)(x - 3) \leq 0$

5. $x^2 + 3x - 10 \leq 0$

6. $x^2 - 2x - 15 \geq 0$

Exercises 7–10. Solve each inequality. Write the solution set in interval notation.

7. $(x + 8)(x + 1)(x - 3) \geq 0$

8. $(q - 1)(q + 4)(3q - 15) \leq 0$

9. $(t^2 - 3)(t + 11) < 0$

10. $(y + 9)(y^3 - 10y) > 0$

Exercises 11–14. Solve each inequality. Write the solution set in interval notation. See Example 2.

11. $\dfrac{x - 3}{x + 10} > 0$

12. $\dfrac{x + 2}{x - 7} \leq 0$

13. $\dfrac{2}{x-2} \ge 3$

14. $\dfrac{1}{x+4} < 5$

C *Exercises 15–16. Use a grapher to check each exercise graphically.*

15. Exercise 7

16. Exercise 12

8.5   Quadratic Functions and Their Graphs

EXAMPLE 1

Match each graph to its equation. Describe how the graph differs from the graph of $f(x) = x^2$.

1.
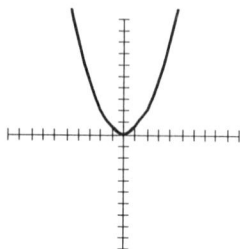

2.
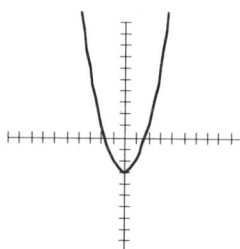

3.
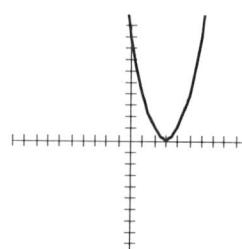

4.
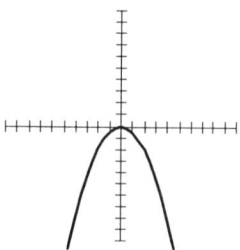

5.
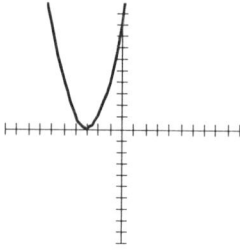

6.
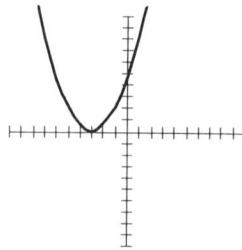

a. $f(x) = -\dfrac{1}{2}x^2$

b. $f(x) = (x+3)^2$

c. $f(x) = x^2 - 3$

d. $f(x) = \dfrac{1}{2}(x+3)^2$

e. $f(x) = \dfrac{1}{2}x^2$

f. $f(x) = (x-3)^2$

Solution: First, sketch the graph of $f(x) = x^2$ for comparison, as shown at the right.

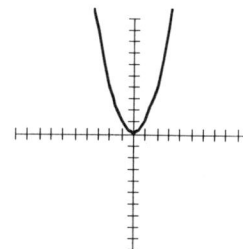

1. This graph is similar to the graph of $f(x) = x^2$. However, it is not shifted but looks wider, so it is most likely of the form $f(x) = ax^2$, corresponding to either graph (a) or (e). Because

the graph opens upward, this must be the graph of $f(x) = \dfrac{1}{2}x^2$, or equation (e).

2. This graph has the same shape as $f(x) = x^2$ but has been shifted downward 3 units. The equation must be of the form $f(x) = x^2 + k$ where $k = -3$. Therefore, this must be the graph of $f(x) = x^2 - 3$, or equation (c).

3. This graph also has the same shape as $f(x) = x^2$ but has been shifted to the right by 3 units. The equation must be of the form $f(x) = (x - h)^2$ where $h = 3$. Therefore, this must be the graph of $f(x) = (x - 3)^2$, or equation (f).

4. This graph is wider than the graph of $f(x) = x^2$ but has not been shifted, and because it opens downward, the equation must be of the form $f(x) = -ax^2$. This must be the graph of $f(x) = -\dfrac{1}{2}x^2$, or equation (a).

5. This graph has the same shape as $f(x) = x^2$ but has been shifted to the left by 3 units. The equation must be of the form $f(x) = (x - h)^2$ where $h = -3$. Therefore, this must be the graph of $f(x) = [x - (-3)]^2 = (x + 3)^2$, or equation (b).

6. This graph is wider than the graph of $f(x) = x^2$ but has also been shifted 3 units to the left. The equation must be of the form $f(x) = a(x - h)^2$, where $h = -3$. Therefore, this must be the graph of $f(x) = \dfrac{1}{2}(x + 3)^2$, or equation (d).

EXERCISES

Exercises 1–4. Find the vertex of each quadratic function.

1. $f(x) = 2x^2$
2. $g(x) = x^2 - 4$
3. $h(x) = (x + 3)^2$
4. $f(x) = 3(x - 2)^2 + 3$

Exercises 5–14. Sketch the graph of each quadratic function. Find the vertex and axis of symmetry.

5. $f(x) = x^2 - 5$
6. $g(x) = x^2 + 4$

7. $h(x) = (x + 4)^2$
8. $f(x) = (x - 2)^2$

9. $g(x) = (x - 1)^2 + 3$
10. $h(x) = (x + 5)^2 - 6$

11. $G(x) = 4x^2$

12. $F(x) = -\dfrac{1}{5}x^2$

13. $H(x) = -(x+1)^2 - 4$

14. $f(x) = \dfrac{1}{3}(x-3)^2 + 5$

C *Exercises 15–18. Use a grapher to graph the first function of each pair below. Then use its graph to predict the graph of the second function. Check your prediction by graphing both on the same set of axes.*

15. $f(x) = x^4$
   $F(x) = x^4 + 6$

16. $g(x) = \sqrt{x}$
   $G(x) = \sqrt{x-6}$

17. $h(x) = (3x-2)^3$
   $H(x) = (3x-2)^3 - 5$

18. $f(x) = \sqrt[3]{x} + 5$
   $F(x) = \sqrt[3]{x-3} + 5$

8.6   Further Graphing of Quadratic Functions

EXAMPLE 1   *Objective 1: Write functions in the form* $y = a(x-h)^2 + k$.

Graph $f(x) = 2x^2 + 6x + 7$. Find the vertex and any intercepts.

Solution: Replace $f(x)$ with y and complete the square on x to write the equation in the form $y = a(x-h)^2 + k$.

$$y = 2x^2 + 6x + 7$$

$$y - 7 = 2x^2 + 6x \qquad \text{Subtract 7 from both sides to isolate the } x\text{-variable terms.}$$

$$y - 7 = 2(x^2 + 3x) \qquad \text{Factor 2 from the terms } 2x^2 + 6x.$$

The coefficient of x is 3. Then $\frac{1}{2}(3) = \frac{3}{2}$ and $\left(\frac{3}{2}\right)^2 = \frac{9}{4}$. Add $\frac{9}{4}$ to the right side inside

parentheses and add $2\left(\frac{9}{4}\right)$ to the left side.

$$y - 7 + 2\left(\frac{9}{4}\right) = 2\left(x^2 + 3x + \frac{9}{4}\right)$$

$$y - \frac{5}{2} = 2\left(x + \frac{3}{2}\right)^2 \qquad \text{Simplify the left side and factor the right side.}$$

$$y = 2\left(x + \frac{3}{2}\right)^2 + \frac{5}{2} \qquad \text{Add } \frac{5}{2} \text{ to both sides.}$$

$$f(x) = 2\left(x + \frac{3}{2}\right)^2 + \frac{5}{2} \qquad \text{Replace } y \text{ with } f(x).$$

Since $a = 2$, the parabola opens upward with vertex $\left(-\frac{3}{2}, \frac{5}{2}\right)$ and axis of symmetry $x = -\frac{3}{2}$.

To find the y-intercept, let $x = 0$ and solve for y. Then

$$f(0) = 2(0)^2 + 6(0) + 7 = 7$$

Thus, the y-intercept is $(0, 7)$.

To find the x-intercepts, notice that the discriminant is $b^2 - 4ac = 36 - 4(2)(7) = -20$ which is negative. This means that the solutions of this quadratic function are both complex; therefore, there are no x-intercepts.

Use this information to sketch the graph of the parabola.

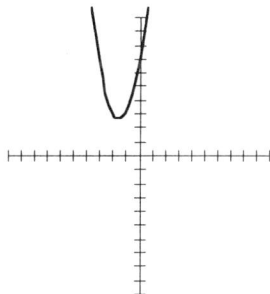

EXAMPLE 2   *Objective 2: Find the vertex of a parabola.*

Find the vertex of the graph of the quadratic function $f(x) = 2x^2 + 20x - 3$.

Solution: First, notice that $a = 2$, $b = 20$, and $c = -3$. Then

$$\frac{-b}{2a} = \frac{-(20)}{2(2)} = \frac{-20}{4} = -5$$

The x-value of the vertex is -5. To find the corresponding $f(x)$ or y-value, find $f(-5)$.

$$f(-5) = 2(-5)^2 + 20(-5) - 3 = -53$$

Thus, the vertex of this parabola is $(-5, -53)$ and the axis of symmetry is $x = -5$.

EXERCISES

Exercises 1–4. Write each quadratic function in the form $y = a(x-h)^2 + k$. Name the vertex. See Example 1.

1. $f(x) = x^2 + 8x + 10$

2. $g(x) = -x^2 + 4x + 5$

3. $h(x) = -2x^2 + 12x + 11$

4. $f(x) = 3x^2 + 15x + 2$

Exercises 5–8. Find the vertex of each quadratic function. See Example 2.

5. $f(x) = 3x^2 + 36x - 1$

6. $g(x) = 2x^2 - 10x + 3$

7. $h(x) = x^2 - 18x - 4$

8. $f(x) = 2x^2 + 56x + 15$

Exercises 9–14. Graph each quadratic function.

9. $g(x) = \frac{1}{2}x^2 + 3x - 1$

10. $h(x) = 2x^2 + 5x + 3$

11. $f(x) = x^2 + 6x - 1$

12. $F(x) = -x^2 - x + 3$

13. $G(x) = \frac{1}{3}x^2 - 2x + 5$

14. $H(x) = -2x^2 + 10x - 3$

15. If a projectile is fired straight upward from the top of a 50-foot tower with an initial speed of 40 feet per second, then its height h in feet after t seconds is given by the equation $h(t) = -16t^2 + 40t + 50$. Find the maximum height of the projectile.

16. The cost C in dollars of manufacturing x compact discs is given by $C(x) = \frac{1}{2}x^2 - 600x + 200,000$.

(a) Find the number of compact discs that must be manufactured to minimize the cost.

(b) Find the minimum cost.

17. Find two positive numbers whose sum is 100 and whose product is as large as possible. [Hint: Let x and $100 - x$ represent the two positive numbers. Their product can be described by the function $f(x) = x(100 - x)$.]

C *Exercises 18–19. Use a grapher to graph each quadratic function and use the zoom and trace features to approximate the coordinates of the vertex. Then check your approximation algebraically.*

18. $f(x) = 3.1x^2 - 5x + 7.9$ **19.** $f(x) = 6.5x^2 + 2.3x - 1.8$

Chapter 8   Hints and Warnings

- To **solve a quadratic equation in x by completing the square:**
    Step 1:   If the coefficient of x^2 is not 1, divide both sides of the equation by the coefficient of x^2.
    Step 2:   Isolate the variable terms.
    Step 3:   Complete the square by adding half of the coefficient of x to both sides.
    Step 4:   Write the resulting trinomial as the square of a binomial.
    Step 5:   Apply the square root property.
- To **solve a quadratic equation of general form using the quadratic formula:**
    Step 1:   Convert the equation into the form $ax^2 + bx + c = 0$.
    Step 2:   Substitute the values of a, b, and c into the formula:
    $$x = \frac{-b \pm \sqrt{b^2 - 4ac}}{2a}$$
- To **solve an equation using quadratic methods,** be sure to analyze the equation to see if substitution would be useful (for instance, if an equation contains a repeated variable expression). Also check to see if one variable term is the square of the other (for instance, x^4 is the square of x^2).
- To **solve a polynomial inequality:**
    Step 1:   Write the inequality in standard form.
    Step 2:   Solve the related equation.
    Step 3:   Use the solutions from Step 2 to separate the number line into intervals.
    Step 4:   Use a test point in each interval to determine whether values in that interval satisfy the original inequality.
    Step 5:   Write the solution set as the union of intervals whose test point value is a solution.
- To **solve a rational inequality:**
    Step 1:   Solve for values that make a denominator 0.
    Step 2:   Solve the related equation.
    Step 3:   Use solutions from Steps 1 and 2 to separate the number line into intervals. *Don't forget to include all values of x which make the denominator equal to 0.*
    Step 4:   Use a test point in each interval to determine whether values in that interval satisfy the original inequality.
    Step 5:   Write the solution set as the union of intervals whose test point value is a solution. *When reporting a solution set for an inequality involving either ≤ or ≥, don't forget that a value which makes the denominator equal to 0 must be excluded from the solution set. That is, in this situation any interval starting or ending with a value that makes the denominator equal to zero must use either (or) to mark the endpoint rather than [or] as would normally be used.*
- To **find the vertex of a parabola,** write the equation in the form $y = ax^2 + bx + c$ and find $\left(\frac{-b}{2a}, f\left(\frac{-b}{2a}\right)\right)$.

CHAPTER 8 Practice Test

Solve each equation for the variable.

1. $3x^2 - 8x = 11$

2. $(x - 3)^2 = 18$

3. $q^2 + 8q - 14 = 0$

4. $2y^2 + 18y + 3 = 0$

5. $5x^2 + 7x + 2 = 0$

6. $b^2 + 2b = 9$

7. $\dfrac{3x}{x - 3} - \dfrac{2}{x + 3} = \dfrac{1}{x^2 - 9}$

8. $a^4 + 2a^2 - 15 = 0$

9. $x^6 - 16x^2 = 32 - 2x^4$

10. $(x - 2)^2 - 10(x - 2) + 9 = 0$

Solve each equation for the variable by completing the square.

11. $x^2 - 2x = -12$

12. $2a^2 - 15 = 8a$

Solve each inequality for x. Write the solution set in interval notation.

13. $3x^2 + 2x < 8$

14. $(x^2 + 3x)(x^2 - 81) > 0$

15. $\dfrac{12x + 3}{x^2 - 25} \le 0$

Graph each function. Find the vertex.

16. $f(x) = \dfrac{1}{4}x^2$

17. $G(x) = 2(x + 3)^2 - 8$

18. $h(x) = x^2 - 8x + 7$

19. $F(x) = -2x^2 + 3x + 5$

20. An 18-foot ladder is leaning against a house. The distance from the bottom of the ladder to the house is 6 feet less than the distance from the top of the ladder to the ground. Find how far the top of the ladder is from the ground. Give an exact answer and a one-decimal-place approximation.

21. Working together, two political campaign workers can stuff envelopes for the candidate's mailing list in 6 hours. Working alone, one worker can finish the same job in 3 hours less time than the other worker. How long does it take each worker to complete the same job if they work alone?

CHAPTER 9

Conic Sections

9.1 The Parabola and the Circle

EXAMPLE 1 *Objective 1: Graph parabolas of the form $x = a(y - k)^2 + h$.*

Graph the parabola $x = \frac{1}{2}(y - 2)^2 - 3$.

Solution: The equation is in the form $x = a(y - k)^2 + h$ with $a = \frac{1}{2}$, $k = 2$, and $h = -3$. Because $a > 0$, the parabola opens to the right. The vertex (h, k) is $(-3, 2)$, and the axis of symmetry is the line $y = 2$. When $y = 0$, $x = -1$, the x-intercept. The parabola is sketched next. Also, a table containing a few ordered pair solutions is shown.

x	y
-1	0
-3	2
5	6

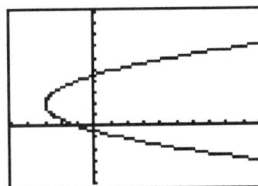

EXAMPLE 2 *Objective 2: Use the distance formula and the midpoint formula.*

Find the distance between and the midpoint of the points $(1, 1)$ and $(-7, -3)$.

Solution: First, use the distance formula. It makes no difference which point we call (x_1, y_1) and which point we call (x_2, y_2). Let $(x_1, y_1) = (1, 1)$ and $(x_2, y_2) = (-7, -3)$.

$$d = \sqrt{(x_2 - x_1)^2 + (y_2 - y_1)^2}$$
$$= \sqrt{(-7 - 1)^2 + (-3 - 1)^2}$$
$$= \sqrt{(-8)^2 + (-4)^2}$$
$$= \sqrt{64 + 16}$$
$$= \sqrt{80}$$
$$= 4\sqrt{5}$$

The distance between the two points is $4\sqrt{5}$.

Second, use the midpoint formula.

$$\text{midpoint} = \left(\frac{x_1 + x_2}{2}, \frac{y_1 + y_2}{2} \right)$$
$$= \left(\frac{1 + (-7)}{2}, \frac{1 + (-3)}{2} \right)$$
$$= \left(\frac{-6}{2}, \frac{-2}{2} \right)$$
$$= (-3, -1)$$

The midpoint of the segment joining the points $(1, 1)$ and $(-7, -3)$ is $(-3, -1)$.

EXAMPLE 3 *Objective 3: Graph circles of the form* $(x-h)^2+(y-k)^2=r^2$.

Graph $(x+1)^2+(y+1)^2=16$.

Solution: The equation can be written as $(x+1)^2+(y+1)^2=4^2$ with $h=-1$, $k=-1$, and $r=4$. The center is $(-1, -1)$, and the radius is 4.

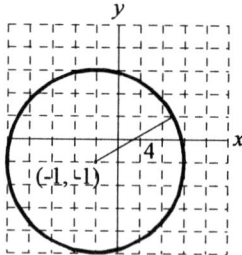

EXERCISES

Exercises 1–4. Find the vertex of and sketch the graph of each parabola. See Example 1.

1. $x=2y^2+4$
2. $x=(y-1)^2-3$
3. $x=(y+4)^2+2$
4. $x=-(y-5)^2+6$

Exercises 5–8. For each pair of points, find the distance between them and find the midpoint of the segment joining them. See Example 2.

5. $(3, 5), (6, 1)$
6. $(2, -1), (0, 4)$
7. $(-7, 2), (3, -8)$
8. $(-5, -3), (-10, 9)$

Exercises 9–14. Find the center and the radius of each circle, and then sketch. See Example 3..

9. $x^2+y^2=1$
10. $x^2+(y+2)^2=9$

11. $(x-1)^2+(y+1)^2=4$
12. $(x+2)^2+(y+1)^2=1$

13. $x^2+2x+y^2=0$
14. $x^2+y^2-4y=0$

Exercises 15–16. Write an equation of the circle with the given center and radius.

15. Center: (5, –3); radius: 5

16. Center: (–1, 10); radius: 7

17. Determine whether the triangle with the vertices (0, 0), (3, 4), and (5, –2) is an isosceles triangle.

18. Determine whether the triangle with vertices (–1, 8), (–3, 0), and (2, 3) is an isosceles triangle.

C *Exercises 19–20. Use a grapher to graph each circle.*

19. $x^2 + y^2 = 48$

20. $2x^2 + 2y^2 - 60 = 0$

9.2   The Ellipse and the Hyperbola

EXAMPLE 1   *Objective 1: Define and graph an ellipse.*

Graph $49x^2 + 4y^2 = 196$.

Solution: Although this equation contains a sum of squared terms in x and y on the same side of the equation, this is not the equation of a circle because the coefficients of x^2 and y^2 are not the same. When this happens, the graph is an ellipse. Because the standard form of the equation of an ellipse has 1 on one side, divide both sides of this equation by 196.

$$49x^2 + 4y^2 = 196$$

$$\frac{49x^2}{196} + \frac{4y^2}{196} = \frac{196}{196} \qquad \text{Divide both sides by 196.}$$

$$\frac{x^2}{4} + \frac{y^2}{49} = 1 \qquad \text{Simplify.}$$

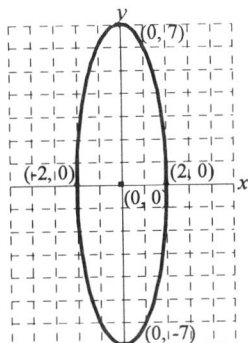

Now recognize that this is the equation of an ellipse with center (0, 0), x-intercepts 2 and –2, and y-intercepts 7 and –7. The sketch of the ellipse is shown.

EXAMPLE 2   *Objective 1: Define and graph an ellipse.*

Graph $\dfrac{(x-1)^2}{16} + \dfrac{(y+2)^2}{4} = 1$.

Solution: This graph has center $(1, -2)$. Also notice that $a = 4$ and $b = 2$. To find four points on the graph of the ellipse, first graph the center $(1, -2)$. Because $a = 4$, count 4 units right and 4 units left of the point with the coordinates $(1, -2)$. Next, because $b = 2$, start at $(1, -2)$ and count 2 units up and 2 units down to find two more points on the ellipse.

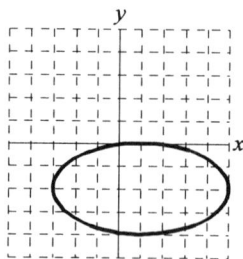

EXAMPLE 3   *Objective 2: Define and graph a hyperbola.*

Graph $9x^2 - 4y^2 = 36$.

Solution: Because this is a difference of squared terms in x and y on the same side of the equation, its graph is a hyperbola, as opposed to an ellipse or a circle. The standard form of the equation of a hyperbola has 1 on one side, so divide both sides of the equation by 36.

$$9x^2 - 4y^2 = 36$$

$$\dfrac{9x^2}{36} - \dfrac{4y^2}{36} = \dfrac{36}{36} \qquad \text{Divide both sides by 36.}$$

$$\dfrac{x^2}{4} - \dfrac{y^2}{9} = 1 \qquad \text{Simplify.}$$

Now recognize that this is the equation of a hyperbola, centered at $(0, 0)$, with $a = 2$ and $b = 3$ and x-intercepts of 2 and -2. The sketch of the hyperbola is shown.

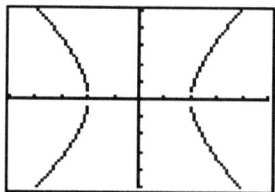

EXERCISES

Exercises 1–4. Sketch the graph of each equation. See Example 1.

1. $\dfrac{x^2}{9} + \dfrac{y^2}{4} = 1$

2. $\dfrac{x^2}{16} + \dfrac{y^2}{25} = 1$

3. $18x^2 + 2y^2 = 18$

4. $25x^2 + 225y^2 = 625$

Exercises 5–6. Sketch the graph of each equation. See Example 2.

5. $\dfrac{(x-3)^2}{4} + \dfrac{(y+2)^2}{1} = 1$

6. $\dfrac{(x+2)^2}{4} + \dfrac{(y-1)^2}{9} = 1$

Exercises 7–10. Sketch the graph of each equation. See Example 3.

7. $\dfrac{x^2}{36} - \dfrac{y^2}{4} = 1$

8. $\dfrac{y^2}{16} - \dfrac{x^2}{9} = 1$

9. $81y^2 - 4x^2 = 324$

10. $25x^2 - y^2 = 25$

Exercises 11–16. Identify whether each equation when graphed will be a parabola, circle, ellipse or hyperbola.

11. $x - y^2 = 4$

12. $x^2 + 4y^2 = 4$

13. $6x^2 + 6y^2 = 6$

14. $(x-2)^2 + 9(y+3)^2 = 9$

15. $(x-1)^2 + (y+5)^2 = 25$

16. $y^2 - x^2 = 16$

C *Exercises 17–18. Use a grapher to graph each ellipse.*

17. $5x^2 + 2y^2 = 10$

18. $3x^2 + 27y^2 = 9$

9.3 Solving Nonlinear Systems of Equations

EXAMPLE 1 *Objective 1: Solve a nonlinear system by substitution.*

Solve the system

$$\begin{cases} x^2 + y^2 = 9 \\ 2x + y = 1 \end{cases}$$

Solution: We can solve this system if we solve one equation for one of the variables. Solving the second equation for y would be a good choice.

$$2x + y = 1 \qquad \text{Second equation.}$$
$$y = 1 - 2x \qquad \text{Solve for } y.$$

Replace y with $1 - 2x$ in the first equation, and then solve for x.

$$x^2 + y^2 = 9 \qquad \text{First equation.}$$
$$x^2 + (1 - 2x)^2 = 9 \qquad \text{Replace } y \text{ with } 1 - 2x.$$
$$x^2 + 1 - 4x + 4x^2 = 9$$
$$5x^2 - 4x - 8 = 0$$
$$a = 5, b = -4, \text{ and } c = -8$$
$$x = \frac{-(-4) \pm \sqrt{(-4)^2 - 4(5)(-8)}}{2(5)}$$
$$x = \frac{2 \pm 2\sqrt{11}}{5}$$

Let $x = \dfrac{2 + 2\sqrt{11}}{5}$ and then let $x = \dfrac{2 - 2\sqrt{11}}{5}$ in the equation $y = 1 - 2x$ to find corresponding y-values.

Let $x = \dfrac{2 + 2\sqrt{11}}{5}$

$$y = 1 - 2x$$
$$y = 1 - 2\left(\frac{2 + 2\sqrt{11}}{5}\right)$$
$$y = 1 - \frac{4 + 4\sqrt{11}}{5}$$
$$y = \frac{1 - 4\sqrt{11}}{5}$$

Let $x = \dfrac{2 - 2\sqrt{11}}{5}$

$$y = 1 - 2x$$
$$y = 1 - 2\left(\frac{2 - 2\sqrt{11}}{5}\right)$$
$$y = 1 - \frac{4 - 4\sqrt{11}}{5}$$
$$y = \frac{1 + 4\sqrt{11}}{5}$$

The solution set is

$$\left\{ \left(\frac{2 + 2\sqrt{11}}{5}, \frac{1 - 4\sqrt{11}}{5} \right), \left(\frac{2 - 2\sqrt{11}}{5}, \frac{1 + 4\sqrt{11}}{5} \right) \right\}. \text{ Check}$$

both solutions in both equations. Both solutions satisfy both equations, so both are solutions of the system. The graph of each equation in the system is shown.

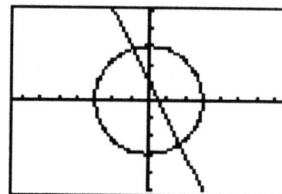

EXAMPLE 2   *Objective 2: Solve a nonlinear system by elimination.*

Solve the system

$$\begin{cases} 4x^2 + 25y^2 = 100 \\ 9x^2 + y^2 = 36 \end{cases}$$

Solution: Use addition or elimination to solve this system. To eliminate y^2 when we add the two equations, multiply both sides of the second equation by –25. Then

$$\begin{cases} 4x^2 + 25y^2 = 100 \\ -25(9x^2 + y^2 = 36) \end{cases} \quad \text{is equivalent to} \quad$$

$$\begin{cases} 4x^2 + 25y^2 = 100 \\ -225x^2 - 25y^2 = -900 \end{cases}$$

$$-221x^2 = -800$$

$$x^2 = \frac{-800}{-221}$$

$$x = \pm\sqrt{\frac{800}{221}}$$

To find corresponding the y-values, let $x = \sqrt{\dfrac{800}{221}}$ and $x = -\sqrt{\dfrac{800}{221}}$ in either original equation. We choose the second equation.

Let $x = \sqrt{\dfrac{800}{221}}$

$$9x^2 + y^2 = 36$$

$$9\left(\sqrt{\frac{800}{221}}\right)^2 + y^2 = 36$$

$$\frac{7200}{221} + y^2 = 36$$

$$y^2 = 36 - \frac{7200}{221}$$

$$y^2 = \frac{7956}{221} - \frac{7200}{221}$$

$$y = \pm\sqrt{\frac{756}{221}}$$

Let $x = -\sqrt{\dfrac{800}{221}}$

$$9x^2 + y^2 = 36$$

$$9\left(-\sqrt{\frac{800}{221}}\right)^2 + y^2 = 36$$

$$\frac{7200}{221} + y^2 = 36$$

$$y^2 = 36 - \frac{7200}{221}$$

$$y^2 = \frac{7956}{221} - \frac{7200}{221}$$

$$y = \pm\sqrt{\frac{756}{221}}$$

The solution set is $\left\{ \left(\sqrt{\dfrac{800}{221}}, \sqrt{\dfrac{756}{221}}\right), \left(\sqrt{\dfrac{800}{221}}, -\sqrt{\dfrac{756}{221}}\right), \left(-\sqrt{\dfrac{800}{221}}, \sqrt{\dfrac{756}{221}}\right), \left(-\sqrt{\dfrac{800}{221}}, -\sqrt{\dfrac{756}{221}}\right) \right\}$.

Check all four ordered pairs in both equations of the system. The graph of each equation in this system is shown.

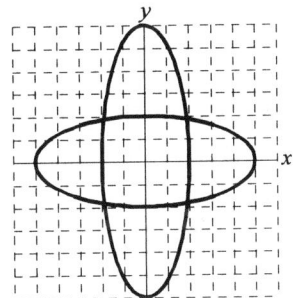

EXERCISES

Exercises 1–8. Solve each system of equations. See Examples 1 and 2.

1. $\begin{cases} y = 2x \\ y = x^2 - x \end{cases}$

2. $\begin{cases} x + 4 = y \\ y = x^2 + 8x + 16 \end{cases}$

3. $\begin{cases} y = 3 \\ 4y^2 - x^2 = 20 \end{cases}$

4. $\begin{cases} 2x - y = 2 \\ 6x - 5 = y + x^2 \end{cases}$

5. $\begin{cases} x^2 + y^2 = 4 \\ y - x^2 = -2 \end{cases}$

6. $\begin{cases} x^2 - y = 3 \\ 2x^2 - y^2 = 6 \end{cases}$

7. $\begin{cases} x^2 + y^2 = 9 \\ x^2 + 9y^2 = 9 \end{cases}$

8. $\begin{cases} x^2 + 3y^2 = 1 \\ 5x^2 - 3y^2 = 1 \end{cases}$

9. The sum of the squares of two numbers is 125. The difference of the squares of the two numbers is 117. Find the two numbers.

10. The sum of the squares of two numbers is 73. The product of the two numbers is 24. Find the two numbers.

C *Exercises 11–12. Use a grapher to solve each system graphically.*

11. $\begin{cases} 8x^2 + 8y^2 = 80 \\ x^2 - 3y^2 = 1 \end{cases}$

12. $\begin{cases} y = 5x^2 + 8x - 1 \\ 0.8x + 3 = y \end{cases}$

9.4   Nonlinear Inequalities and Systems of Inequalities

EXAMPLE 1   *Objective 1: Sketch the graph of a nonlinear inequality.*

Graph $3(x-1)^2 + 3(y+2)^2 < 27$.

Solution: First, recognize that the graph of the related equation is a circle. The related equation is equivalent to the equation $(x-1)^2 + (y+2)^2 = 9$, which has center $(1, -2)$ and radius 3. Graph the circle with a dotted line. To determine which region of the plane contains the solutions, select a test point in either region and determine whether the coordinates of the point satisfy the inequality. We choose $(0, 0)$ as a test point.

$$3(x-1)^2 + 3(y+2)^2 < 27$$
$$3(0-1)^2 + 3(0+2)^2 < 27 \qquad \text{Let } x = 0 \text{ and } y = 0.$$
$$3 + 3(4) < 27$$
$$15 < 27 \qquad \text{True.}$$

Because this statement is true, the solution set is the region containing $(0, 0)$; that is, the points inside of, but not on, the circle, as shaded in the figure.

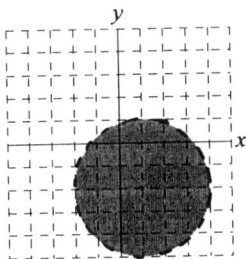

EXAMPLE 2   *Objective 2: Sketch the solution set of a system of nonlinear inequalities.*

Graph the system

$$\begin{cases} y \ge \dfrac{1}{2}x^2 \\ y - 2x \le 5 \end{cases}$$

Solution: Graph each inequality on the same set of axes. The intersection is the shaded region along with its boundary lines. The coordinates of the points of intersection can be found by solving the related system

$$\begin{cases} y = \dfrac{1}{2}x^2 \\ y - 2x = 5 \end{cases}$$

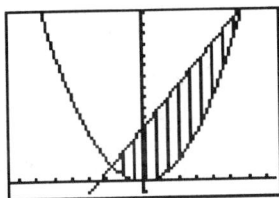

EXERCISES

Exercises 1–6. Graph each inequality. See Example 1.

1. $y \le x^2 + 5x - 3$

2. $y \le -x^2 + 6$

3. $(x + 3)^2 + (y - 2)^2 < 4$

4. $x^2 + y^2 \ge 36$

5. $\dfrac{x^2}{9} + y^2 < 1$

6. $y^2 - 3x^2 \le 1$

Exercises 7–10. Graph the solution of each system. See Example 2.

7. $\begin{cases} y \le x^2 \\ y \ge -3 \end{cases}$

8. $\begin{cases} y \le -x^2 \\ y \ge -3 \end{cases}$

9. $\begin{cases} x^2 + y^2 \le 16 \\ y \ge x - 2 \end{cases}$

10. $\begin{cases} 4x^2 + 25y^2 \le 100 \\ 9x^2 + y^2 \le 36 \end{cases}$

Chapter 9   Hints and Warnings

- To **find the distance** d **between two points** (x_1, y_1) and (x_2, y_2), use the formula
  $d = \sqrt{(x_2 - x_1)^2 + (y_2 - y_1)^2}$.

- To **find the midpoint of the line segment** whose end points are (x_1, y_1) and (x_2, y_2), use the coordinates
  $\left(\dfrac{x_1 + x_2}{2}, \dfrac{y_1 + y_2}{2} \right)$.

- A **parabola** is represented by the equations $y = a(x - h)^2 + k$ or $x = a(y - k)^2 + h$. *Remember that graphs of the* $y = a(x - h)^2 + k$ *equations open along the y-axis and the graphs of the* $x = a(y - k)^2 + h$ *equations open along the x-axis. In* $y = a(x - h)^2 + k$, *if a > 0, the parabola opens upward; if a < 0, the parabola opens downward. In* $x = a(y - k)^2 + h$, *if a > 0, the parabola opens to the right; if a < 0, the parabola opens to the left. In either equation, the vertex of the parabola is given by (h, k).*

- A **circle** is represented by the equation $(x-h)^2 + (y-k)^2 = r^2$. *Remember that (h, k) is the center of the circle and that r represents the length of the radius of the circle. If a circle has an equation $x^2 + y^2 = r^2$, the circle has center (0, 0).*

- An **ellipse** is represented by the equation $\dfrac{x^2}{a^2} + \dfrac{y^2}{b^2} = 1$. *Remember that this ellipse is centered at (0, 0) and that the x-intercepts are a and –a and the y-intercepts are b and –b. If an ellipse has the equation $\dfrac{(x-h)^2}{a^2} + \dfrac{(y-k)^2}{b^2} = 1$, the ellipse is centered at the point (h, k).*

- A **hyperbola** is represented by the equations $\dfrac{x^2}{a^2} - \dfrac{y^2}{b^2} = 1$ or $\dfrac{y^2}{b^2} - \dfrac{x^2}{a^2} = 1$. *Remember that such a hyperbola has its center at (0, 0). In $\dfrac{x^2}{a^2} - \dfrac{y^2}{b^2} = 1$, the hyperbola opens along the x-axis and has x-intercepts at a and –a. In $\dfrac{y^2}{b^2} - \dfrac{x^2}{a^2} = 1$, the hyperbola opens along the y-axis and has y-intercepts at b and –b.*

- To **solve a nonlinear system of equations,** either the substitution method or the addition (elimination) method may be used.

CHAPTER 9   Practice Test

Find the distance between each pair of points.

1. $(4, -9), (-1, 3)$

2. $(0, 8), (-4, -2)$

3. $\left(\sqrt{3}, 2\sqrt{5}\right), \left(-3\sqrt{3}, 3\sqrt{5}\right)$

Find the midpoint of the line segment having the given pair of end points.

4. $(6, -3), (4, -5)$

5. $\left(\dfrac{3}{4}, -\dfrac{1}{5}\right), \left(\dfrac{1}{2}, -\dfrac{3}{5}\right)$

Sketch the graph of each equation. Find the center, radius, vertices, intercepts, and/or asymptotes as appropriate.

6. $x^2 + y^2 = 9$

7. $x^2 - y^2 = 9$

8. $4x^2 + 16y^2 = 64$

9. $y = -(x-3)^2 + 6$

10. $x^2 + y^2 + 10y = 9$

11. $x = y^2 + 4y + 2$

12. $\dfrac{(x+2)^2}{4} + \dfrac{(y+1)^2}{9} = 1$

13. $y^2 - x^2 = 16$

Solve each system.

14. $\begin{cases} x^2 + y^2 = 100 \\ 8y = 6x \end{cases}$

15. $\begin{cases} x^2 + y^2 = 98 \\ x^2 - y^2 = 64 \end{cases}$

16. $\begin{cases} y = -x^2 + 4x + 4 \\ y = 4 \end{cases}$

Graph the solution of each system of inequalities.

17. $\begin{cases} y \geq \dfrac{1}{2}x - 3 \\ y \leq -x^2 - 2x + 5 \end{cases}$

18. $\begin{cases} x^2 + \dfrac{y^2}{9} \leq 1 \\ y \leq -x \end{cases}$

19. $\begin{cases} y^2 - x^2 \leq 16 \\ y \geq 0.25x - 1 \end{cases}$

20. Which graph most resembles the graph of $\dfrac{(x-h)^2}{a^2} + \dfrac{(y-k)^2}{b^2} = 1$ if $h > 0$, $k < 0$, and $a < b$?

A.
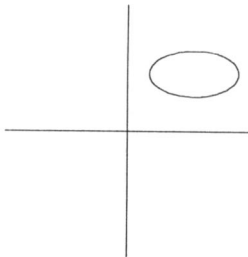

B.
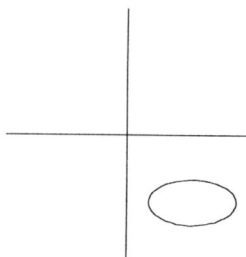

C.
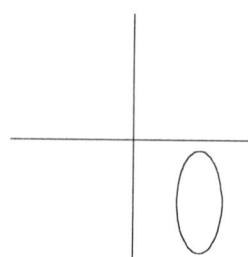

D.
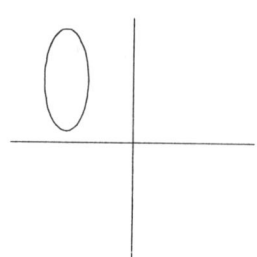

21. A bridge has an arch in the shape of half of an ellipse. If the equation of the ellipse measured in feet is $169x^2 + 256y^2 = 43{,}264$, find the height of the arch from the road at the center of the arch and the width of the arch at the base.

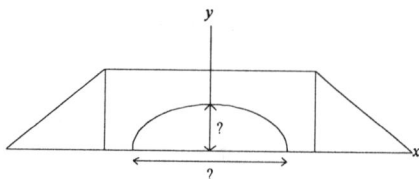

CHAPTER 10

Exponential and Logarithmic Functions

10.1   Inverse Functions

EXAMPLE 1   *Objective 1: Compose functions.*

If $f(x) = (2x^2 + 1)^2$ and $g(x) = \sqrt{x}$, find each of the following compositions.

a. $(f \circ g)(4)$ **b.** $(g \circ f)(4)$ **c.** $(f \circ g)(x)$ **d.** $(g \circ f)(x)$

Solution: a.   Recall that $(f \circ g)(x) = f(g(x))$.

$$(f \circ g)(4) = f(g(4))$$
$$= f(2) \qquad \text{Because } g(4) = \sqrt{4} = 2.$$
$$= [2(2)^2 + 1]^2$$
$$= 81$$

b. $(g \circ f)(4) = g(f(4))$
$$= g(1089) \qquad \text{Because } f(4) = [2(4)^2 + 1]^2 = 33^2 = 1089.$$
$$= \sqrt{1089}$$
$$= 33$$

c. $(f \circ g)(x) = f(g(x))$
$$= f(\sqrt{x}) \qquad \text{Replace } g(x) \text{ with } \sqrt{x}.$$
$$= [2(\sqrt{x})^2 + 1]^2$$
$$= (2x + 1)^2$$
$$= 4x^2 + 4x + 1 \qquad \text{Simplify.}$$

d. $(g \circ f)(x) = g(f(x))$
$$= g([2x^2 + 1]^2) \qquad \text{Replace } f(x) \text{ with } (2x^2 + 1)^2.$$
$$= \sqrt{(2x^2 + 1)^2}$$
$$= 2x^2 + 1$$

EXAMPLE 2   *Objective 3: Use the horizontal line test to test whether a function is a one-to-one function.*

Determine whether each graph is the graph of a one-to-one function.

a.   **b.**   **c.**

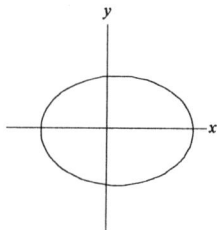   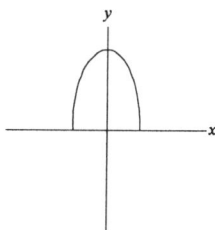   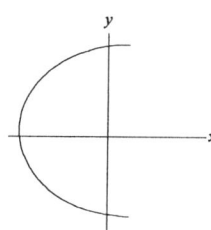

Solution: **a.** This graph does not pass the horizontal line test and is not the graph of a function. Thus, this is not the graph of a one-to-one function.

        b. This graph is a function, but it does not pass the horizontal line test. Thus, this is not the graph of a one-to-one function.

        c. This graph passes the horizontal line test but it does not pass the vertical line test for a function. Thus, this is not the graph of a one-to-one function.

EXAMPLE 3   *Objective 5: Find the equation of the inverse of a function.*

    Find the equation of the inverse of $f(x) = \sqrt[3]{2-x}$.

Solution: First, let $y = f(x)$.

$$f(x) = \sqrt[3]{2-x}$$
$$y = \sqrt[3]{2-x}$$

Next, interchange x and y and solve for y.

$$x = \sqrt[3]{2-y}$$ Interchange x and y.

$$x^3 = \left(\sqrt[3]{2-y}\right)^3$$ Cube both sides of the equation.

$$x^3 = 2 - y$$

$$y = 2 - x^3$$ Solve for y.

Thus, $f^{-1}(x) = 2 - x^3$.

EXERCISES

Exercises 1–8. Let $f(x) = 4x^2$, $g(x) = 5 + \sqrt{x}$, and $h(x) = 10x + 3$. Find the following. See Example 1.

1. $(f \circ g)(x)$ **2.** $(f \circ h)(x)$ **3.** $(g \circ h)(x)$ **4.** $(g \circ f)(x)$

5. $(h \circ f)(x)$ **6.** $(h \circ g)(x)$ **7.** $(g \circ h)(9)$ **8.** $(g \circ f)(-3)$

Exercises 9–12. Determine whether each list is a one-to-one function.

9. $\{(3, 4), (5, -1), (2, 0), (3, 6), (7, 3)\}$ **10.** $\{(1, 3), (2, 2), (3, 0), (4, -1), (5, 2)\}$

11.

State (Input)	Alaska	Arkansas	Delaware	Idaho	Maine
Area Code (Output)	907	501	302	208	207

(data from U.S. Department of Commerce)

12.

College (Input)	Clarke College	Adrian College	Gordon College	Kenyon College	Lyndon State College
1994–1995 Enrollment (Output)	1002	1059	1178	1510	1178

(data from Peterson's Guides, © 1995)

Exercises 13–16. Determine whether each function is a one-to-one function. See Example 3.

13. $f(x) = 2x^2 + 3x$

14. $g(x) = x^3 + 2x - 1$

15. $h(x) = x^3 + 4x^2 + x - 5$

16. $F(x) = 16x - 5$

Exercises 17–20. Each function is one-to-one. Find the inverse of each function. See Example 4.

17. $f(x) = 5 - 3x$

18. $g(x) = \dfrac{x + 2}{5}$

19. $h(x) = \dfrac{x - 3}{x}$

20. $F(x) = 4x^2 - 3$, $x \geq 0$

21. If $f(x) = \dfrac{1}{2}x - 7$, show that $f^{-1}(x) = 2x + 14$.

22. If $f(x) = 2x^3 + 1$, show that $f^{-1}(x) = \sqrt[3]{\dfrac{x - 1}{2}}$.

C *Exercises 23–24. Find the inverse of each given one-to-one function. Then use a grapher to graph the function and its inverse in a square window.*

23. $f(x) = \dfrac{1}{10}x - 7$

24. $g(x) = \sqrt[3]{4x - 7}$

10.2   Exponential Functions

EXAMPLE 1   *Objective 2: Graph exponential functions.*

Graph the exponential functions $f(x) = \left(\dfrac{3}{10}\right)^x$ and $g(x) = \left(\dfrac{10}{3}\right)^x$ on the same set of axes.

Solution: Graph each function by plotting points. Set up a table of values for each of the two functions. Then plot the points and connect them with a smooth curve.

$f(x) = \left(\dfrac{3}{10}\right)^x$ $g(x) = \left(\dfrac{10}{3}\right)^x$

x	$f(x)$
0	1
1	0.3
2	0.09
3	0.027
−1	3.333
−2	11.11

x	$g(x)$
0	1
1	3.333
2	11.11
3	37.04
−1	0.3
−2	0.09

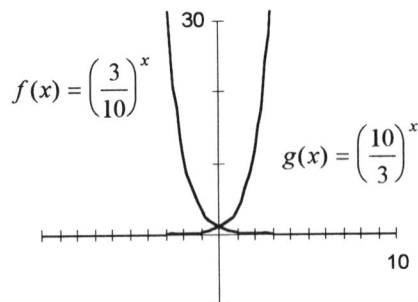

Each function is one-to-one. The *y*-intercept of both functions is 1. Notice that for *f*, whose base is between 0 and 1, the graph is decreasing. For *g*, whose base is greater than 1, the graph is increasing.

EXAMPLE 2   *Objective 3: Solve equations of the form $b^x = b^y$.*

Solve each equation for *x*.

a. $5^x = 25$ **b.** $2^{3x-1} = 32$ **c.** $16^{2x-3} = 64^{5x-2}$

Solution: **a.** First, write 25 as a power of 5 and then use the uniqueness of b^x to solve.

$$5^x = 25$$
$$5^x = 5^2$$

Because the bases are the same and nonnegative, by the uniqueness of b^x, we have then that the exponents are equal. Thus,
$$x = 2$$

b. First, write 32 as a power of 2.

$$2^{3x-1} = 32$$
$$2^{3x-1} = 2^5$$

$3x - 1 = 5$ Apply the uniqueness of b^x.

$\quad 3x = 6$ Add 1 to both sides.

$\quad\quad x = 2$ Divide both sides by 3.

c. First, write both 16 and 64 as powers of 4.

$$16^{2x-3} = 64^{5x-2}$$
$$(4^2)^{2x-3} = (4^3)^{5x-2}$$
$$4^{4x-6} = 4^{15x-6}$$

$4x - 6 = 15x - 6$ Apply the uniqueness of b^x.

$\quad 0 = 11x$ Add 6 to both sides and subtract 4*x* from both sides.

$\quad x = 0$ Divide both sides by 11.

EXERCISES

Exercises 1–4. Graph each exponential function. See Example 1.

1. $y = 2^x - 5$

2. $y = \left(\dfrac{2}{5}\right)^x$

3. $f(x) = 3 - 2^x$

4. $f(x) = 3 - \left(\dfrac{1}{2}\right)^x$

Exercises 5–12. Solve each equation. See Example 2.

5. $32^x = 4$

6. $36^x = 216$

7. $5^{2x+1} = 125$

8. $4 = 64^{x-3}$

9. $125^{x+1} = 625^{2x-1}$

10. $0.1^x = 100^{x+3}$

11. $27^{2x-8} = 9^{x-4}$

12. $11^{3x-4} = 121^{5-x}$

13. Over the period 1985–1992, the population of the country Togo was increasing at a rate of 3.7% annually. There were approximately 3,900,000 residents in 1992. Assuming that the growth pattern remains the same, predict the number of residents living in Togo after 5 years. Use $y = 3,900,000(2.7)^{0.037t}$. (data from The World Bank)

14. Over the period 1985–1992, the population of Kuwait was decreasing at a rate of 2.3% annually. There were approximately 1,400,000 residents in 1992. Assuming that this decline continues, predict the number of residents living in Kuwait after 8 years. Use $y = 1,400,000(2.7)^{-0.023t}$. (data from The World Bank)

15. Suppose you deposit $5000 in a savings account that pays 4% interest that is compounded quarterly. If no other deposits or withdrawals are made, what is the value of the account after 4 years? Use $A = P\left(1 + \dfrac{r}{n}\right)^{nt}$.

16. Suppose you take out a loan for $8000 at a rate of 9% interest that is compounded monthly. What amount will you owe at the end of 2 years? Use $A = P\left(1 + \dfrac{r}{n}\right)^{nt}$.

C *Exercises 17–18. Use a grapher to estimate the value of $12,000 invested at 9.5% interest compounded daily after the given numbers of years. Use $A = P\left(1 + \dfrac{r}{n}\right)^{nt}$. Use the fact that there are 365 days in a year.*

17. 2 years   **18.** 7 years

10.3   Logarithmic Functions

EXAMPLE 1   *Objective 2: Solve logarithmic equations by using exponential notation.*

Solve each equation for x.
  a. $\log_2 64 = x$ **b.** $\log_4 x = -2$ **c.** $\log_x 81 = 4$

Solution: **a.** $\log_2 64 = x$ means $2^x = 64$. Solve $2^x = 64$ for x.

$$2^x = 64$$
$$2^x = 2^6$$
$$x = 6$$

The solution set is {6}. To check, see that $\log_2 64 = 6$ since $2^6 = 64$.

  b. $\log_4 x = -2$ means $4^{-2} = x$ or

$$x = \frac{1}{16}$$

The solution set is $\left\{ \dfrac{1}{16} \right\}$.

  c. $\log_x 81 = 4$ means $x^4 = 81$ and $x > 0$ and $x \neq 1$.

$$x = 3$$

Even though $(-3)^4 = 81$, the base b of a logarithm must be positive. Thus, the solution set is {3}.

EXAMPLE 2   *Objective 2: Solve logarithmic equations by using exponential notation.*

Simplify.
  a. $\log_{10} 1$ **b.** $\log_5 5^{-8}$ **c.** $4^{\log_4 12}$

Solution: **a.**   Because $\log_b 1 = 0$, $\log_{10} 1 = 0$.

  b.   Because $\log_b b^x = x$, $\log_5 5^{-8} = -8$.

  c.   Because $b^{\log_b x} = x$, $4^{\log_4 12} = 12$.

EXAMPLE 3   *Objective 3: Identify and graph logarithmic functions.*

Graph the logarithmic function $y = \log_4 x$.

Solution: Write the equation using exponential notation as $4^y = x$. Find some ordered pair solutions that satisfy this equation. Plot the points and connect them with a smooth curve. The domain of this function is $(0, \infty)$ and the range of the function is all real numbers.

x	y
1	0
4	1
0.25	−1

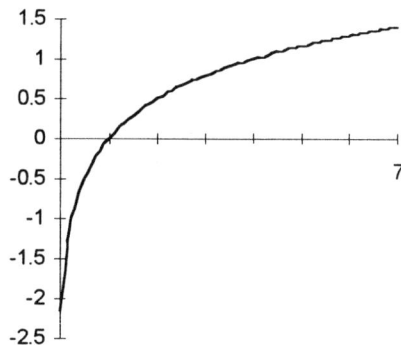

EXERCISES

Exercises 1–6. Find the value of each logarithmic expression.

1. $\log_{10} 1000$

2. $\log_6 = 216$

3. $\log_3 \dfrac{1}{27}$

4. $\log_5 \dfrac{1}{25}$

5. $\log_{49} 7$

6. $\log_9 27$

Exercises 7–12. Solve each equation. See Example 1.

7. $\log_7 49 = x$

8. $\log_{1/2} 8 = x$

9. $\log_{16} x = \dfrac{1}{2}$

10. $\log_{10} x = -1$

11. $\log_x 32 = 5$

12. $\log_x 8 = -3$

Exercises 13–16. Simplify. See Example 2.

13. $\log_{11} 11^3$

14. $\log_7 7^{-4}$

15. $\log_3 1$

16. $9^{\log_9 2}$

Exercises 17–20. Graph each logarithmic function. See Example 3.

17. $y = \log_8 x$

18. $y = \log_{1/8} x$

19. $y = \log_{1/10} x$

20. $y = \log_{10} x$

10.4   Properties of Logarithms

EXAMPLE 1   *Objective 1: Apply the product property of logarithms.*

Use the product property to simplify. Assume that variables represent positive numbers.
a. $\log_7 4 + \log_7 5$ **b.** $\log_3 (y-3) + \log_3 (y+3)$

Solution: a. $\log_7 4 + \log_7 5 = \log_7 (4 \cdot 5) = \log_7 20$

b. $\log_3 (y-3) + \log_3 (y+3) = \log_3[(y-3)(y+3)] = \log_3 (y^2 - 9)$

EXAMPLE 2   *Objective 2: Apply the quotient property of logarithms.*

Use the quotient property to simplify. Assume that variables represent positive numbers.
a. $\log_6 56 - \log_6 7$ **b.** $\log_2 (x+8) - \log_2 y$

Solution: a. $\log_6 56 - \log_6 7 = \log_6 \dfrac{56}{7} = \log_6 8$

b. $\log_2 (x+8) - \log_2 y = \log_2 \dfrac{x+8}{y}$

EXAMPLE 3   *Objective 3: Apply the power property of logarithms.*

Use the power property to simplify. Assume that variables represent positive numbers.
a. $\log_{10} x^5$ **b.** $\log_7 \sqrt[6]{y}$

Solution: a. $\log_{10} x^5 = 5 \log_{10} x$

b. $\log_7 \sqrt[6]{y} = \dfrac{1}{6} \log_7 y$

EXERCISES

Exercises 1–4. Write each as the logarithm of a single expression. Assume that all variables represent positive numbers. See Example 1.

1. $\log_2 12 + \log_2 6$

2. $\log_5 \dfrac{1}{2} + \log_5 (4x-8)$

3. $\log_3 y + \log_3 5 + \log_3 (y-1)$

4. $\log_4 (y+2) + \log_4 (y-3) + \log_4 7$

Exercises 5–8. Write each as the logarithm of a single expression. Assume that all variables represent positive numbers. See Example 2.

5. $\log_{10} 36 - \log_{10} 9$

6. $\log_5 15 - \log_5 \dfrac{1}{3}$

7. $\log_8 20 + \log_8 3 - \log_8 4$

8. $\log_3(7x + 21) - \log_3 7$

Exercises 9–12. Rewrite each logarithm using the power rule. Assume that all variables represent positive numbers. See Example 3.

9. $3\log_3 y$

10. $-4\log_{10} 3$

11. $-\log_4 x$

12. $\dfrac{1}{2}\log_9 36$

Exercises 13–16. Write each as the logarithm of a single expression. Assume that all variables represent positive numbers.

13. $\dfrac{1}{3}\log_7 27 + \log_7 x$

14. $\log_4 10 - 2\log_4 5$

15. $3\log_2 x + \log_2 6 - 2\log_2 x$

16. $4\log_5 x - 3\log_5 z - 5\log_5 y + \log_5 2$

Exercises 17–20. Write each expression as a sum or difference of multiples of logarithms. Assume that variables represent positive numbers.

17. $\log_{10}\dfrac{8}{x}$

18. $\log_5 2y^3$

19. $\log_3\dfrac{x^5}{y}$

20. $\log_6\left(\dfrac{a}{b}\right)^2$

10.5   Common Logarithms, Natural Logarithms, and Change of Base

EXAMPLE 1   *Objective 2: Evaluate common logarithms of powers of 10.*

Solve $\log(6x + 1) = 2.1$. Give an exact solution, and then approximate the solution to four decimal places.

Solution: Write the logarithmic equation using exponential notation. Keep in mind that the base of a common logarithm is understood to be 10.

$$\log(6x + 1) = 2.1$$

$$10^{2.1} = 6x + 1 \qquad \text{Write using exponential notation.}$$

$$10^{2.1} - 1 = 6x$$

$$x = \frac{10^{2.1} - 1}{6} \qquad \text{Exact solution.}$$

To approximate, use a calculator. To four decimal places, $x \approx 20.8154$.

EXAMPLE 2   *Objective 4: Evaluate natural logarithms of powers of e.*

Solve $\ln(10x^2) = 14$. Give an exact and an approximate solution (to four decimal places).

Solution: Write the equation using exponential notation. Keep in mind that the base of a natural logarithm is understood to be e.

$$\ln(10x^2) = 14$$

$$e^{14} = 10x^2 \qquad \text{Write using exponential notation.}$$

$$\frac{e^{14}}{10} = x^2$$

$$\pm\sqrt{\frac{e^{14}}{10}} = x \qquad \text{Exact solution.}$$

To approximate, use a calculator. To four decimal places, $x \approx \pm 346.7859$.

EXAMPLE 3   *Objective 5: Apply the change of base formula.*

Approximate $\log_7 18$ to four decimal places.

Solution: Use the change of base property to write $\log_7 18$ as a quotient of logarithms to base 10.

$$\log_7 18 = \frac{\log_{10} 18}{\log_{10} 7} \qquad \text{Use the change of base property.}$$

$$\approx \frac{1.255272505}{0.84509804} \qquad \text{Approximate logarithms by calculator.}$$

$$\approx 1.485357255 \qquad \text{Simplify by calculator.}$$

EXERCISES

Exercises 1–4. Use a calculator to approximate each logarithm to four decimal places.

1. $\log 12$ **2.** $\log 6.067$ **3.** $\ln 10$ **4.** $\ln 4.75$

Exercises 5–8. Find the exact value.

5. $\log 100{,}000$ **6.** $\log 0.01$ **7.** $\ln e^{-5}$ **8.** $\ln \sqrt{e}$

Exercises 9–16. Solve each equation. Give an exact solution and a four-decimal-place approximation. See Examples 1 and 2.

9. $\log x = -1.15$ **10.** $\ln x = 3.6$ **11.** $\ln 2x = 7.1$ **12.** $\log \frac{1}{2}x = 1.89$

13. $\log(3x + 5) = 2.34$ **14.** $\log(x - 6) = -1.24$ **15.** $\ln(5x - 12) = 0.81$ **16.** $\ln(8 - 3x) = 3.62$

Exercises 17–20. Approximate each logarithm to four decimal places. See Example 3.

17. $\log_2 9$ **18.** $\log_7 16$ **19.** $\log_{1/4} 10$ **20.** $\log_5 \dfrac{1}{8}$

Exercises 21–22. Use the formula $R = \log\left(\dfrac{a}{T}\right) + B$ to find the intensity R on the Richter scale of the earthquakes fitting the descriptions given. Approximate the solution to the nearest tenth.

21. Amplitude a is 300 micrometers, time T between waves is 2.1 seconds, and B is 4.2.

22. Amplitude a is 250 micrometers, time T between waves is 3.8 seconds, and B is 2.6.

C *Exercises 23–24. Use a grapher to graph each function.*

23. $f(x) = e^{2x} - 4$ **24.** $f(x) = e^{2-x}$

10.6   Exponential and Logarithmic Equations and Problem Solving

EXAMPLE 1   *Objective 1: Solve exponential equations.*

Solve $2e^{2x-3} = 6$.

Solution: To solve, use the logarithm property of equality and take the logarithm of both sides. For this example, we use the natural logarithm.

$$2e^{2x-3} = 6$$

$$e^{2x-3} = 3 \qquad \text{Divide both sides by 2.}$$

$$\ln e^{2x-3} = \ln 3 \qquad \text{Take the natural log of both sides.}$$

$$2x - 3 = \ln 3 \qquad \text{Apply the property } \ln_b b^x = x.$$

$$2x = \ln 3 + 3 \qquad \text{Add 3 to both sides.}$$

$$x = \frac{\ln 3 + 3}{2} \qquad \text{Exact solution.}$$

The exact solution is $\dfrac{\ln 3 + 3}{2}$. If a decimal approximation is preferred,

$\dfrac{\ln 3 + 3}{2} \approx \dfrac{4.098612289}{2} \approx 2.0493$, to four decimal places. The solution set is $\left\{\dfrac{\ln 3 + 3}{2}\right\}$ or approximately $\{2.0493\}$.

EXAMPLE 2   *Objective 2: Solve logarithmic equations..*

Solve $4 + \log_2(3x - 1) = 7$

Solution: Begin by combining like terms.

$$4 + \log_2(3x - 1) = 7$$
$$\log_2(3x - 1) = 3$$

Next, write the equation using exponential notation and solve for x.

$$\log_2(3x - 1) = 3$$

$2^3 = 3x - 1$	Write using exponential notation.
$8 = 3x - 1$	Simplify.
$9 = 3x$	Add 1 to both sides.
$3 = x$	Divide both sides by 3.

The solution set is $\{3\}$.

EXERCISES

Exercises 1–6. Solve each equation. Give an exact solution, and also approximate the solution to four decimal places. See Example 1.

1. $5^x = 60$

2. $e^x = 15$

3. $e^{4x} = 16$

4. $4^{3x} = 9$

5. $2^{3x-4} = 10$

6. $e^{5x+1} = 24$

Exercises 7–14. Solve each equation. See Example 2.

7. $\log_5 6 + \log_5 x = 0$

8. $\log_7 4 + 2\log_7 x = 0$

9. $\log_4 8 - 3\log_4 x = 3$

10. $\log_3 18 - \log_3 x = 2$

11. $\log_7(4x + 5) = 2$

12. $4\log_9 x - 3\log_9 x = 3$

13. $\log_2(x^2 - 11x + 60) = 5$

14. $\log_5 x + \log_5(4x - 15) = 2$

15. Over the period 1985–1992, the population of Sudan was increasing at a rate of 2.8% annually. There were 26,587,000 residents in 1992. Assuming that the growth pattern remains the same, predict the number of residents living in Sudan after 7 years. Use $y = y_0 e^{0.028t}$. (data from The World Bank)

16. Over the period 1985–1992, the population of France was increasing at a rate of 0.6% annually. There were 57,338,000 residents in 1992. Assuming that the growth pattern remains the same, predict the number of residents living in France after 10 years. Use $y = y_0 e^{0.006t}$. (data from The World Bank)

17. Find how long it takes $3000 to double if it is invested at 3% interest compounded quarterly. Use $A = P\left(1+\dfrac{r}{n}\right)^{nt}$.

18. How long will it take $5000 to earn $1500 in interest if it is invested at 8% interest compounded monthly? Use $A = P\left(1+\dfrac{r}{n}\right)^{nt}$.

19. Find the expected height of a boy who would normally weigh 95 pounds. Use $w = 0.00185h^{2.67}$.

20. Find the average atmospheric pressure of Quito, Ecuador, which is 9222 feet above sea level. Use $P = 14.7e^{-0.21x}$, where P is the average atmospheric pressure in pounds per square inch and x is altitude above sea level in miles. (Note: 1 mile = 5280 feet). (data from Defense Mapping Agency)

C *Exercises 21–22. Use a grapher to solve each equation graphically. Round solutions to two decimal places.*

21. $e^{2.6x} = 18$ **22.** $\ln(4x-1) = 2x-3$

Chapter 10   Hints and Warnings

- To **test for a one-to-one function,** graph the function and see if every horizontal line intersects the graph at most once. If so, the function is a one-to-one function.
- To **find the inverse of a one-to-one function** $f(x)$:
  Step 1: Replace $f(x)$ with y.
  Step 2: Interchange x and y.
  Step 3: Solve for y.
  Step 4: Replace y with $f^{-1}(x)$.
- Remember that if b is a real number and $b > 0$, $b \neq 1$, then $\log_b 1 = 0$, $\log_b b^x = x$, and $b^{\log_b x} = x$.
- Let x, y, and b be positive numbers, $b \neq 1$. Remember that $\log_b xy = \log_b x + \log_b y$ (product property), $\log_b \dfrac{x}{y} = \log_b x - \log_b y$ (quotient property), and $\log_b x^r = r\log_b x$ (power property). *A common mistake made*

with the quotient property is to confuse $\dfrac{\log_b x}{\log_b y}$ *with* $\log_b \dfrac{x}{y}$. *Don't forget that* $\dfrac{\log_b x}{\log_b y} \neq \log_b x - \log_b y$ *and*

that $\log_b \dfrac{x}{y} = \log_b x - \log_b y$ *is the proper statement of the quotient property of logarithms.*

- Remember that the **common logarithm** $\log x$ means $\log_{10} x$. The **natural logarithm** $\ln x$ means $\log_e x$.
- To **solve exponential equations,** use the logarithm property of equality which says that $\log_b a = \log_b c$ is equivalent to $a = c$ if $\log_b a$ and $\log_b c$ are real numbers, $b \neq 1$. *That is, to solve an exponential equation of the form* $a^x = b$, *you can transform this equation using the logarithm property of equality (along with the power property) to* $x \log a = \log b$. *Then you need only solve for x in the usual way.*
- To **solve logarithmic equations,** first simplify the equation (if necessary) using the properties of logarithms and then rewrite using exponential notation.

CHAPTER 10 Practice Test

If $f(x) = 2x + 3$, *and* $g(x) = x^2 - 1$, *find the following.*

1. $(f \circ g)(0)$ **2.** $(f \circ g)(x)$ **3.** $(g \circ f)(x)$

On the same set of axes, graph the given one-to-one function and its inverse.

4. $f(x) = \dfrac{1}{3}x + 8$

Determine whether the given graph is the graph of a one-to-one function.

5.

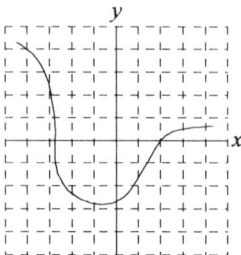

6.

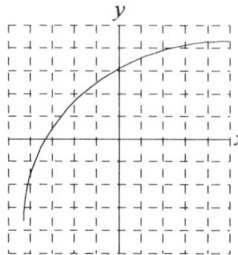

Determine whether each function is one-to-one. If it is one-to-one, find an equation or a set of ordered pairs that defines the inverse function of the given function.

7. $y = 9x^2 + 1$ **8.** $f = \{(10, -3), (11, 2), (12, -1), (13, -4)\}$

9.

State (Input)	Maryland	Texas	Ohio	Nevada	Louisiana
Order of entry in U.S. (Output)	7	28	17	36	18

Use the properties of logarithms to write each expression as a single logarithm.

10. $\log_5 72 - \log_5 8$ **11.** $2\log_3 x + \log_3 x - \log_3(2x - 1)$

12. Write the expression $\log_2 10x^5y$ as the sum or difference of multiples of logarithms.

13. If $\log_b 32 = 7.18$ and $\log_b 6 = 3.7$, find the value of $\log_b\left(\dfrac{216}{32}\right)$.

14. Approximate $\log_3 15$ to four decimal places.   **15.** Solve $12^{3x-1} = 144$ for x. Give an exact solution.

16. Solve $8^{2x-1} = 30$ for x. Give an exact solution, and approximate the solution to four decimal places.

Solve each logarithmic equation for x. Give an exact solution.

17. $\log_8 x = 3$ **18.** $\ln\sqrt[3]{e^2} = x$ **19.** $\log_4(14x + 60) = 4$

20. $\log_3 8 + \log_3 x = 3$ **21.** $\log_2(11x - 1) - \log_2(x - 1) = 4$

22. Solve $\ln(2x - 9) = 6$ for x, accurate to four decimal places.

23. Graph $y = \left(\dfrac{1}{3}\right)^x + 3$.   **24.** Graph the functions $y = 2^x$ and $y = \log_2 x$ on the same set of axes.

Use the formula $A = P\left(1 + \dfrac{r}{n}\right)^{nt}$ to solve Exercises 25 and 26.

25. Find the amount in the account if $12,000 is invested for 9 years at 7% compounded quarterly.

26. Find how long it will take $4000 to grow to $6000 if the money is invested at 6% interest compounded monthly.

Use the population growth model $y = y_0 e^{kt}$ to solve Exercises 27 and 28.

27. Over the period 1985–1992, the population of Guam was increasing at a rate of 3.2% annually. There were 150,000 residents in 1992. Assuming that the growth pattern remains the same, predict the number of residents living in Guam after 20 years. (data from The World Bank)

28. A population of rabbits has an annual growth rate of 16.1%. If there are currently 50 rabbits, how long will it take to reach 180 rabbits?

C 29. Use a grapher to approximate the solution of $e^{3x-1} = 5 - 2x$ to two decimal places.

CHAPTER 11

Sequences, Series, and the Binomial Theorem

11.1   Sequences

EXAMPLE 1   *Objective 1: Write the terms of a sequence given its general terms.*

Let the general term of a sequence be given by $a_n = 100 - \dfrac{2n^2}{5}$. Find

a.   the first four terms of this sequence   b. a_{20} c.   the fortieth term

Solution: a.   Evaluate a_n when n is 1, 2, 3, and 4.

$$a_n = 100 - \frac{2n^2}{5}$$

$$a_1 = 100 - \frac{2(1)^2}{5} = 99\frac{3}{5}$$

$$a_2 = 100 - \frac{2(2)^2}{5} = 98\frac{2}{5}$$

$$a_3 = 100 - \frac{2(3)^2}{5} = 96\frac{2}{5}$$

$$a_4 = 100 - \frac{2(4)^2}{5} = 93\frac{3}{5}$$

b.   Evaluate a_n for $n = 20$.

$$a_{20} = 100 - \frac{2(20)^2}{5} = -60$$

c.   Evaluate a_n for $n = 40$.

$$a_{40} = 100 - \frac{2(40)^2}{5} = -540$$

EXAMPLE 2   *Objective 2: Find the general term of a sequence.*

Find the general term a_n for the sequence whose first few terms are $\dfrac{1}{2}, \dfrac{2}{3}, \dfrac{3}{4}, \dfrac{4}{5}, \dfrac{5}{6} \cdots$.

Solution: For each term, the numerator and denominator is 1 more than the numerator and denominator of the previous term. Additionally, in each term the denominator is 1 more than the numerator. A general term for the sequence might be $a_n = \dfrac{n}{n+1}$. We can check this by finding the first several terms this sequence.

$$a_n = \frac{n}{n+1} \qquad a_2 = \frac{2}{2+1} = \frac{2}{3} \qquad a_4 = \frac{4}{4+1} = \frac{4}{5}$$

$$a_1 = \frac{1}{1+1} = \frac{1}{2} \qquad a_3 = \frac{3}{3+1} = \frac{3}{4} \qquad a_5 = \frac{5}{5+1} = \frac{5}{6}$$

It appears that $a_n = \dfrac{n}{n+1}$ adequately describes the pattern of the given terms of the sequence.

EXERCISES

Exercises 1–8. Write the first four terms of each sequence whose general term is given. See Example 1.

1. $a_n = n + 10$ **2.** $a_n = -\dfrac{n}{2}$ **3.** $a_n = 3n - 1$ **4.** $a_n = \dfrac{2}{n}$

5. $a_n = (-1)^n (n + 2)$ **6.** $a_n = \dfrac{n+2}{n}$ **7.** $a_n = 3^n$ **8.** $a_n = 10 - n^2$

Exercises 9–12. Find the indicated term for each sequence whose general term is given.

9. $a_n = \dfrac{n-2}{n+4}$; a_{18} **10.** $a_n = 2^{n-1}$; a_7 **11.** $a_n = (-1)^n (3n - 2)$; a_{11} **12.** $a_n = 100 - 2n^2$; a_{15}

Exercises 13–16. Find a general term a_n for each sequence whose first four terms are given. See Example 2.

13. $6, 7, 8, 9, \ldots$ **14.** $1, \dfrac{1}{4}, \dfrac{1}{9}, \dfrac{1}{16}, \ldots$

15. $7, 10, 13, 16, \ldots$ **16.** $4, -8, 16, -32, \ldots$

17. A new restaurant had 200 customers in its first week of business. The manager expects the number of customers served to increase by 80 customers each week. Find the number of customers for each of 6 weeks, starting with the first week.

18. A construction firm estimates that one of its new bulldozers will depreciate $12,500 per year. The bulldozer originally cost $100,000. Write an equation of a sequence that describes the value of the bulldozer at the end of each year. Find the value of the bulldozer after 8 years.

C *Exercises 19–20. Use a calculator to find the first four terms of each sequence. Round each term to four decimal places.*

19. $a_n = \sqrt{2n - 1}$ **20.** $a_n = \dfrac{\ln n}{n}$

11.2   Arithmetic and Geometric Sequences

EXAMPLE 1   *Objective 1: Identify arithmetic sequences and their common differences.*

Find the fifteenth term of the arithmetic sequence whose first three terms are $-5, -2, 1$.

Solution: Because the sequence is arithmetic, the fifteenth term is
$$a_{15} = a_1 + (15-1)d = a_1 + 14d$$

We know a_1 is the first term of the sequence, so $a_1 = -5$. Also, d is the constant difference of terms, so $d = a_2 - a_1 = -2 - (-5) = 3$. Thus
$$a_{15} = a_1 + 14d$$
$$= -5 + 14 \cdot 3$$
$$= 37$$

EXAMPLE 2   *Objective 1: Identify arithmetic sequences and their common differences.*

If the fifth term of an arithmetic progression is 10 and the twenty-first term is 18, find the twelfth term.

Solution: We need to find a_1 and d to write the general term, which then enables us to find a_{12}, the twelfth term. The given facts about a_5 and a_{21} lead to a system of linear equations.
$$\begin{cases} 10 = a_1 + 4d \\ 18 = a_1 + 20d \end{cases}$$

Next, solve this system by addition. Multiply both sides of the first equation by -1 so that the system becomes

$$\begin{cases} -10 = -a_1 - 4d \\ 18 = a_1 + 20d \end{cases} \quad \text{and} \quad \begin{cases} -10 = -a_1 - 4d \\ \underline{18 = a_1 + 20d} \\ 8 = 16d \\ \dfrac{1}{2} = d \end{cases}$$

To find a_1, let $d = \dfrac{1}{2}$ in $10 = a_1 + 4d$. Then
$$10 = a_1 + 4d$$
$$10 = a_1 + 4\left(\frac{1}{2}\right)$$
$$10 = a_1 + 2$$
$$8 = a_1$$

Thus, $a_1 = 8$ and $d = \dfrac{1}{2}$, so
$$a_n = 8 + (n-1)\left(\frac{1}{2}\right)$$
$$a_{12} = 8 + (12-1)\left(\frac{1}{2}\right)$$
$$a_{12} = 8 + \frac{11}{2} = \frac{27}{2} = 13\frac{1}{2}$$

EXAMPLE 3   *Objective 2: Identify geometric sequences and their common ratios.*

Find the sixth term of the geometric progression whose first three terms are 14, 21, 31.5.

Solution: Because the sequence is geometric, the sixth term must be $a_1 r^{6-1}$ or $14r^5$. We know that r is the common ratio of terms, so r must be $\dfrac{21}{14}$ or $\dfrac{3}{2}$. Thus

$$a_6 = 14r^5$$

$$a_6 = 14\left(\frac{3}{2}\right)^5 = \frac{1701}{16}$$

EXERCISES

Exercises 1–4. Write the first four terms of the arithmetic or geometric sequence whose first term a_1 and common difference d or common ratio r are given.

1. $a_1 = -16; d = 5$ **2.** $a_1 = 30; d = -8$ **3.** $a_1 = 10; r = 4$ **4.** $a_1 = -4; r = -\dfrac{1}{2}$

Exercises 5–10. Find the indicated term of each sequence.

5. The tenth term of the arithmetic sequence whose first term is 5 and whose common difference is 12.

6. The twelfth term of the arithmetic sequence 20, 17, 14, . . .

7. The fifth term of the geometric sequence 1000, 200, 40, . . .

8. The sixth term of the geometric sequence whose first term is 3 and whose common ratio is –10.

9. If the second term of a geometric progression is 3 and the third term is 12, find a_1 and r.

10. The sixteenth term of the arithmetic sequence whose fifth term is 18 and whose thirteenth term is 24.

Exercises 11–14. Given are the first three terms of a sequence that is either arithmetic or geometric. Based on these three terms, if a sequence is arithmetic find d. If the sequence is geometric, find r.

11. 3, 15, 75   **12.** –7, –5, –3   **13.** 6, 26, 46   **14.** 10, –20, 40

15. You are offered a job with an annual salary of $26,000 with a guarantee of a 4% annual raise for each of the next 5 years of employment. What will your annual salary be during your sixth year of employment? (Note: A 4% annual raise corresponds to a factor of 1.04.)

16. You are offered a job with an annual salary of $26,000 with a guarantee of a $1100 annual raise for each of the next 5 years of employment. What will your annual salary be during your sixth year of employment?

17. The number of online/hybrid computer services was 8,000,000 in 1995, expected to be 13,000,000 in 1996, expected to be 18,000,000 in 1997, and so on. Find the general term of this arithmetic sequence and the expected number of online/hybrid computer services in 2000 if this trend continues as expected. (data from Morgan Stanley)

18. From 1991 through 1995, the sales of gospel music were growing faster than any other major type of popular music. In 1991, sales of gospel music were approximately $298,000,000, and since then sales have been growing at a rate of 22% per year. What were the sales of gospel music in 1995? Predict the level of sales of gospel music for 1999. (data from Gospel Music Association and the Recording Industry Association of America)

11.3   Series

EXAMPLE 1   *Objective 2: Use summation notation.*

Write the series using summation notation.
$7 + 13 + 19 + 25 + 31 + 37$

Solution: Because the difference of each term and the preceding term is 6, the terms correspond to the first six terms of an arithmetic sequence with $a_1 = 7$ and $d = 6$. So $a_n = 7 + (n-1)6 = 7 + 6n - 6 = 1 + 6n$. Thus, using summation notation, we have

$$7 + 13 + 19 + 25 + 31 + 37 = \sum_{i=1}^{6} (1 + 6i)$$

EXAMPLE 2   *Objective 2: Find partial sums.*

Find the sum of the first four terms of the sequence whose general term is $a_n = \dfrac{n+1}{n}$.

Solution: $S_4 = \sum_{i=1}^{4} \dfrac{i+1}{i} = \dfrac{1+1}{1} + \dfrac{2+1}{2} + \dfrac{3+1}{3} + \dfrac{4+1}{4} = 2 + \dfrac{3}{2} + \dfrac{4}{3} + \dfrac{5}{4} = 6\dfrac{1}{12}$.

EXERCISES

Exercises 1–6. Evaluate.

1. $\sum_{i=1}^{6} (i - 5)$

2. $\sum_{i=1}^{3} (i + 2)^2$

3. $\sum_{i=1}^{5} \dfrac{i}{5}$

4. $\sum_{i=1}^{4} (2i - 2)$

5. $\sum_{i=2}^{5} i(i + 3)$

6. $\sum_{i=3}^{6} 2^i$

Exercises 7–12. Write each series using summation notation. See Example 1.

7. $3 + 15 + 75 + 375$

8. $-7 + -5 + -3 + -1 + 1$

9. $6 + 26 + 46 + 66$

10. $10 + -20 + 40$

11. $5 + 10 + 20 + 40 + 80$

12. $2 + 4 + 6 + 8 + 10 + 12 + 14$

13. Find the sum of the first four terms of the sequence whose general term is $a_n = 2n^2$.

14. Find the sum of the first three terms of the sequence whose general term is $a_n = (7n - 5)^3$.

15. A quilter is making a quilt with a pyramid pattern having 14 quilt blocks on the bottom row, 13 quilt blocks in the next row, 12 quilt blocks in the next row, and so on for 14 rows. Write the sequence describing the number of quilt blocks in each row. Find the total number of quilt blocks the quilter will need.

16. You are offered a job with an annual salary of $26,000 with a guarantee of a 4% annual raise for each of the next 5 years of employment. How much will you have earned in total at the end of 4 years of employment? (Note: A 4% annual raise corresponds to a factor of 1.04.)

11.4   Partial Sums of Arithmetic and Geometric Sequences

EXAMPLE 1   *Objective 1: Find the partial sum of an arithmetic sequence.*

Use the partial sum formula to find the sum of the first five terms of the arithmetic sequence. 7, 13, 19, 25, 31, . . .

Solution: Use the formula for S_n of an arithmetic sequence, replacing n with 5, a_1 with 7, and a_n with 31.

$$S_n = \frac{n}{2}(a_1 + a_n) = \frac{5}{2}(7 + 31) = \frac{5}{2}(38) = 95$$

EXAMPLE 2   *Objective 2: Find partial sums of a geometric sequence.*

Use the partial sum formula to find the sum of the first four terms of the geometric sequence. 80, 60, 45, 33.75, . . .

Solution: Use the formula for the partial sum S_n of the terms of a geometric sequence. Here, $n = 4$, the first term $a_1 = 80$, and the common ratio $r = 0.75$.

$$S_n = \frac{a_1(1 - r^n)}{1 - r} = \frac{80[1 - (0.75)^4]}{1 - 0.75} = 218.75$$

EXAMPLE 3   *Objective 3: Find the infinite series of a geometric sequence.*

Find the sum of the terms of the geometric sequence 80, 60, 45, 33.75, . . .

Solution: For this geometric sequence, $r = 0.75$. Because $|r| < 1$, we may use the formula S_∞ of a geometric sequence with $a_1 = 80$ and common ratio $r = 0.75$.

$$S_\infty = \frac{a_1}{1-r} = \frac{80}{1-0.75} = 320$$

EXERCISES

Exercises 1–8. Use the appropriate partial sum formula to find the partial sum of the given arithmetic or geometric sequence. See Examples 1 and 2.

1. Find the sum of the first ten terms of the sequence 10, 15, 20, . . .

2. Find the sum of the first five terms of the sequence 8, 32, 128, . . .

3. Find the sum of the first six terms of the sequence –16, –8, –4, . . .

4. Find the sum of the first six terms of the sequence –10, –8, –6, . . .

5. Find the sum of the first eight terms of the sequence $\frac{1}{2}, \frac{3}{4}, 1 \ldots$

6. Find the sum of the first 7 terms of the sequence 100, 10, 1, . . .

7. Find the sum of the first 25 positive integers.   **8.** Find the sum of the first 100 positive integers.

Exercises 9–14. Find the sum of the terms of each infinite geometric sequence. See Example 3.

9. $1, \frac{7}{8}, \frac{49}{64}, \ldots$ **10.** 49, 42, 36, . . .   **11.** 512, 64, 8, . . .

12. $5, \frac{5}{4}, \frac{5}{16}, \ldots$ **13.** $-7, -\frac{7}{3}, -\frac{7}{9}, \ldots$ **14.** $-10, -6, -\frac{18}{5}, \ldots$

15. During the test of a new antibiotic, the drug killed 4000 cells during the first minute, 2000 cells during the second minute, 1000 cells during the third minute, and so on. Find the total number of cells killed during the first 10 minutes.

11.5   The Binomial Theorem

EXAMPLE 1   *Objective 2: Evaluate factorials.*

Evaluate $\dfrac{8!}{4!6!}$.

Solution: $\dfrac{8!}{4!6!} = \dfrac{8 \cdot 7 \cdot 6 \cdot 5 \cdot 4 \cdot 3 \cdot 2 \cdot 1}{4 \cdot 3 \cdot 2 \cdot 1 \cdot 6 \cdot 5 \cdot 4 \cdot 3 \cdot 2 \cdot 1} = \dfrac{8 \cdot 7}{4 \cdot 3 \cdot 2 \cdot 1} = \dfrac{7}{3}$

EXAMPLE 2   *Objective 3: Use the binomial theorem to expand binomials.*

Use the binomial theorem to expand $(5s - 3t)^6$.

Solution: Let $a = 5s$ and $b = -3t$ in the binomial formula.

$$(5s - 3t)^6 = (5s)^6 + \frac{6}{1!}(5s)^5(-3t) + \frac{6 \cdot 5}{2!}(5s)^4(-3t)^2 + \frac{6 \cdot 5 \cdot 4}{3!}(5s)^3(-3t)^3 + \frac{6 \cdot 5 \cdot 4 \cdot 3}{4!}(5s)^2(-3t)^4$$

$$+ \frac{6 \cdot 5 \cdot 4 \cdot 3 \cdot 2}{5!}(5s)(-3t)^5 + \frac{6 \cdot 5 \cdot 4 \cdot 3 \cdot 2 \cdot 1}{6!}(-3t)^6$$

$$= 15{,}625s^6 - 56{,}250s^5t + 84{,}375s^4t^2 - 67{,}500s^3t^3 + 30{,}375s^2t^4 - 7290st^5 + 729t^6$$

EXAMPLE 3   *Objective 4: Find the nth term in the expansion of a binomial raised to a positive power.*

Find the sixth term of the expansion of $(x + 4)^8$.

Solution: Use the formula, with $n = 8$, $a = x$, $b = 4$, and $r + 1 = 6$. Notice that because $r + 1 = 6$, $r = 5$.

$$\frac{n!}{r!(n-r)!}a^{n-r}b^r = \frac{8!}{5!\,3!}x^{8-5}4^5$$

$$= 56x^3(1024)$$

$$= 57{,}344x^3$$

EXERCISES

Exercises 1–2. Use Pascal's triangle to expand the binomial.

1. $(z - y)^4$

2. $(s + t)^8$

Exercises 3–4. Evaluate each expression. Example 1.

3. $\dfrac{4!}{7!}$

4. $\dfrac{12!}{9!\,3!}$

Exercises 5–10. Use the binomial formula to expand each binomial. See Example 2.

5. $(w + z)^4$

6. $(m - n)^6$

7. $(a-3b)^4$

8. $(2x+y)^5$

9. $(2q+3r)^3$

10. $(5-2p)^4$

Exercises 11–14. Find the indicated term. See Example 3.

11. The seventh term of the expansion of $(q+2r)^{11}$.

12. The third term of the expansion of $(3x-y)^6$.

13. The fifth term of the expansion of $(w-7)^8$.

14. The tenth term of the expansion of $(10a+b)^9$.

Chapter 11   Hints and Warnings

- To **find the general term** a_n **of an arithmetic sequence,** use the formula $a_n = a_1 + (n-1)d$, where a_1 is the first term of the sequence and d is the common difference. *If the common difference is negative, remember to attach the negative sign to the value for d.*

- To **find the general term** a_n **of a geometric sequence,** use the formula $a_n = a_1 r^{n-1}$ where a_1 is the first term of the sequence and r is the common ratio. *If the common ratio is negative, remember to attach the negative sign to the value for r.*

- To **find the partial sum** S_n **of the first** n **terms of an arithmetic sequence,** use the formula

  $S_n = \dfrac{n}{2}(a_1 + a_n)$, where a_1 is the first term of the sequence and a_n is the nth term.

- To **find the partial sum** S_n **of the first** n **terms of a geometric sequence,** use the formula $S_n = \dfrac{a_1(1-r^n)}{1-r}$, where a_1 is the first term of the sequence and r is the common ratio, $r \ne 1$.

- To **find the sum of the terms of an infinite geometric sequence,** use the formula $S_\infty = \dfrac{a_1}{1-r}$, where a_1 is the first term of the sequence and r is the common ratio, $|r| < 1$. *Remember that if* $|r| \ge 1$, S_∞ *does not exist.*

- To **expand a binomial raised to a positive integer power** n**,** use the Binomial Theorem which states that $(a+b)^n = a^n + \dfrac{n}{1!}a^{n-1}b^1 + \dfrac{n(n-1)}{2!}a^{n-2}b^2 + \dfrac{n(n-1)(n-2)}{3!}a^{n-3}b^3 + \cdots + b^n$. *If either of the terms a or b is negative, remember to include the negative sign when substituting for a or b in the Binomial Theorem formula.*

CHAPTER 11 Practice Test

Find the indicated term(s) of the given sequence.

1. The first four terms of the sequence $a_n = \dfrac{n^2}{2n-1}$

2. The first four terms of the sequence $a_n = (n-1)(n+2)$

3. The fiftieth term of the sequence $a_n = 22 + 6(n-1)$

4. The one-hundredth term of the sequence $a_n = (81-2n)^2$

5. The general term of the sequence $\dfrac{6}{7}, \dfrac{12}{7}, \dfrac{24}{7}, \ldots$

6. The general term of the sequence $40, 24, 8, \ldots$

Find the partial sum of the given sequence.

7. S_{16} of the sequence $a_n = 11 + (n-1)5$

8. S_5 of the sequence $a_n = 8(3)^{n-1}$

9. S_∞ of the sequence $a_1 = 10$ and $r = \dfrac{4}{9}$

10. S_∞ of the sequence $2, \dfrac{4}{5}, \dfrac{8}{25}, \ldots$

11. $\displaystyle\sum_{i=1}^{5} (i^2 - 4)$

12. $\displaystyle\sum_{i=4}^{8} (2i-1)(i+2)$

Expand the binomial using Pascal's triangle.

13. $(2a-b)^3$

14. $(y+3)^5$

Expand the binomial using the binomial formula.

15. $(y+3r)^4$

16. $(x-y)^6$

17. The population of a small town is growing yearly according to the sequence defined by $a_n = 800 + 20(n-1)$. Predict the population after 6 years.

18. A gardener is planting shrubs to fill a plot of land in the shape of a trapezoid with 15 shrubs in the first row, 17 shrubs in the second row, 19 shrubs in the third row, and so on for 10 rows. Write the finite series of this sequence and find the total number of shrubs planted.

Practice Final Exam #1

Determine whether each statement is true or false.

1. $16 - 7 = |16 - 7|$

2. All real numbers are rational numbers.

Simplify.

3. $(7 - 4)^3 - |-9 - 8|^2$

4. $13 - 24 \div 3(4)$

5. $\dfrac{3(15 - 13)^3 + (-6)}{(-2)(-3)}$

Evaluate each expression when $x = -2, y = 1, z = 4$.

6. $x^3 + z^2$

7. $\dfrac{xz - y^2}{xy}$

8. Write the statement using mathematical symbols: twice the difference of q and 6 is less than 10.

9. Name the property that is illustrated: $8z - 24 = 8(z - 3)$

10. Write an expression for the total amount of money (in cents) in n nickels and q quarters.

Solve each equation.

11. $8x + 26 = 3x - 4$

12. $\dfrac{h}{5} + \dfrac{h}{10} = 60$

13. $|8 - 3x| = 4$

Solve each equation for the specified variable.

14. $9x - 32y = 11;\ y$

15. $V = lwh;\ h$

Solve each inequality.

16. $-9 + \dfrac{x}{5} \geq 4$

17. $11 \leq 5x + 2 < 18$

18. $|2x - 7| > 14$

19. Find 76% of 180.

20. Find the amount of money in an account after 4 years if a principal of \$10,000 was invested at 6.1% interest compounded quarterly and no withdrawals were made. (Round to the nearest cent.)

21. Name the quadrant in which each point is located: $(-4, 7)$, $(-2, -8)$, $(14, 1)$, and $(3, -6)$

Graph each line.

22. $x - 4y = 16$

23. $-3x + y = 4$

24. $x + y + 6 = 0$

25. $y = 2$

Find an equation of each line satisfying the conditions given.

26. Vertical line through $(8, -3)$

27. Perpendicular to $x = 3$; through $(-2, -6)$

28. Through $(3, 5)$; slope -2

29. Through $(5, 6)$; parallel to $4x + 2y = -5$

30. Find the domain and range of the relation: $\{(0, 9), (3, 2), (-2, 5), (3, -4), (4, 2)\}$. Is the relation a function?

Evaluate each determinant.

31. $\begin{vmatrix} 6 & -4 \\ 2 & -3 \end{vmatrix}$

32. $\begin{vmatrix} 2 & 3 & 1 \\ -3 & 5 & 3 \\ 4 & 0 & -2 \end{vmatrix}$

Solve each system of equations by the addition method or the substitution method.

33. $\begin{cases} x + 2y = 3 \\ 4x - 3y = 12 \end{cases}$

34. $\begin{cases} x + 3z = 0 \\ \dfrac{1}{5}x + 2y + 3z = 0 \\ 4x = 20 \end{cases}$

Solve each system using Cramer's Rule.

35. $\begin{cases} 3x + y = -1 \\ 2x - 5y = 5 \end{cases}$

36. $\begin{cases} 2x - y = 5 \\ 3x + 4y = 2 \end{cases}$

Solve each system using matrices.

37. $\begin{cases} 2x - 5y = 12 \\ x + 4y = 6 \end{cases}$

38. $\begin{cases} x + y + z = 1 \\ 2x + z = 4 \\ -x + 2y + 4z = 3 \end{cases}$

39. A coffee shop mixes its own house blend from Brazilian Rain Forest Coffee and Blue Mountain Dark Roast Coffee. How many pounds of Brazilian Rain Forest Coffee (costing $6 per pound) and Blue Mountain Dark Roast Coffee (costing $10 per pound) are needed to make 30 pounds of the house blend which sells for $9 per pound?

40. A motel in Miami, Florida, charges $39 including tax per night for a room on a weekend. During the week, rooms cost $65 including tax per night. If the total receipts for one week were $115,180 and a total of 2020 rooms were rented, how many rooms were rented during the week and how many were rented on the weekend?

Simplify. Use positive exponents to write answers.

41. $(2xy)^{-5}$

42. $\dfrac{4m^5 n^6}{16m^{-2} n^9}$

43. Write in scientific notation: $47,200$

44. Write without exponents: 6.71×10^{-3}

Perform the indicated operations.

45. $(5x^2 - 6x + 8) - (2x^2 + 5x - 4)$

46. $(5x - 8)(5x + 8)$

Factor each polynomial completely.

47. $9z^3 - 3z^2 - 21z$

48. $x^2 + 11x + 28$

49. Solve for the variable: $(4q + 1)(4q - 1) = 15q^2 + 8$

50. Graph: $y = x^2 - 3x - 10$

51. Find the domain of the rational function: $f(x) = \dfrac{14x^2}{3x - 20}$

52. Write the rational expression in lowest terms: $\dfrac{4x - 2x^2}{18x^2 - 30x}$

Perform the indicated operation.

53. $\dfrac{y^2}{y^2 - 25} \cdot \dfrac{y + 5}{8y}$

54. $\dfrac{14x}{x^2 + 8x + 12} - \dfrac{3}{x + 6}$

Divide. Simplify each answer.

55. $\dfrac{\dfrac{4}{w} + \dfrac{7}{3w}}{\dfrac{7}{2w} - \dfrac{3}{w}}$

56. $\dfrac{25a^2b^2c + 10a^2bc - 45bc^3}{5abc^2}$

57. Use synthetic division to divide $3x^4 - 5x^3 + 7x^2 - 80x - 400$ by $x - 5$.

Solve each equation for x.

58. $\dfrac{3x + 10}{x - 5} = 4$

59. $\dfrac{x}{x + 5} - \dfrac{2}{x - 8} = 1$

60. Suppose that R is inversely proportional to w. If $R = 240$ and $w = 25$, find w when $R = 80$.

Raise to the power or take the root. Assume that all variables represent positive numbers. Write using only positive exponents.

61. $-\sqrt[3]{x^{48}}$

62. $\left(\dfrac{1}{512}\right)^{-1/3}$

63. Rationalize the denominator: $\sqrt{\dfrac{64}{5y^2}}$

64. Perform the indicated operations. Assume that all variables represent positive numbers: $\sqrt{125y^3} - 2y\sqrt{20y}$

65. Perform the indicated operations: $\left(\sqrt{2} + 8\right)\left(\sqrt{3} - 6\right)$

66. Solve the equation for x: $x = \sqrt{9x - 14}$

Perform the indicated operation and simplify. Write the result in the form $a + bi$.

67. $(12 - 5i) - (6 - 8i)$

68. $(8 + 2i)(8 - 2i)$

Solve each equation for the variable.

69. $(x + 4)^2 = 32$

70. $y^2 + 20y + 6 = 0$

71. $b^2 + 5b = 14$

72. $a^4 + 7a^2 - 18 = 0$

73. $(x + 4)^2 - 8(x + 4) + 15 = 0$

74. Solve the inequality for x, and write the solution set in interval notation: $x^2 - 3x < 10$

75. Graph the function, and find the vertex: $h(x) = -x^2 - 4x + 21$

76. A 14-foot ladder is leaning against a house. The distance from the bottom of the ladder to the house is 7 feet less than the distance from the top of the ladder to the ground. Find how far the top of the ladder is from the ground. Give an exact answer and a one-decimal place approximation.

Find the distance between each pair of points, and find the midpoint of the segment joining each pair.

77. (–1, 8), (3, –4)

78. (2, 3), (6, –7)

Sketch the graph of each equation. Find the center, radius, vertices, intercepts, and/or asymptotes as appropriate.

79. $x^2 + y^2 = 16$

80. $\dfrac{x^2}{64} - \dfrac{y^2}{25} = 1$

81. $25x^2 + 9y^2 = 225$

82. $x = (y - 2)^2 - 5$

83. Solve the system: $\begin{cases} y = 2x^2 + 9x + 6 \\ y = 6 \end{cases}$

84. Graph the solution of the system: $\begin{cases} 4x^2 + 16y^2 \le 64 \\ y \le 0.75x + 2 \end{cases}$

85. If $f(x) = 3x - 1$, and $g(x) = 2x^2$, find (a) $(g \circ f)(0)$ and (b) $(f \circ g)(x)$.

86. Find the inverse of the one-to-one function: $f(x) = -2x + 11$

87. Is the function one-to-one? $f = \{(8, -2), (3, 2), (5, -3), (4, -6)\}$

88. Use the properties of logarithms to write the expression as a single logarithm: $3\log_3 x - \log_3 x + \log_3(4x + 3)$

89. Write the expression $\log_5 8x^4z^2$ as the sum or difference of multiples of logarithms.

90. Solve $6^{2x-3} = 216$ for x. Give an exact solution.

91. Solve the logarithmic equation for x: $\log_4 12 + \log_4 x = 2$. Give an exact solution.

C 92. Use a grapher to approximate the solution of $e^{2x+5} = 7 - 3x$ to two decimal places.

Find the indicated term(s) of the given sequence.

93. The first four terms of the sequence $a_n = \dfrac{n^3}{2n^2 + 5}$

94. The fiftieth term of the sequence $a_n = 30 + 25(n - 1)$

95. The general term of the sequence $100, 67, 34, \ldots$

Find the partial sum of the given sequence.

96. S_4 of the sequence $a_n = 7(2)^{n+1}$

97. S_∞ of the sequence $a_1 = 24$ and $r = \dfrac{1}{2}$

98. $\displaystyle\sum_{i=1}^{5} (2i^2 + 2)$

99. Expand the binomial using the binomial formula: $(r - s)^5$

100. A gardener is planting shrubs to fill a plot of land in the shape of a trapezoid with 8 shrubs in the first row, 11 shrubs in the second row, 14 shrubs in the third row, and so on for 7 rows. Write the finite series of this sequence and find the total number of shrubs planted.

Practice Final Exam #2

1. Is the following statement true or false? $2 - 13 = |2 - 13|$
   - **(a)** True
   - **(b)** False
   - **(c)** Not enough information

2. Is the following statement true or false? All rational numbers are real numbers.
   - **(a)** True
   - **(b)** False
   - **(c)** Not enough information

3. Simplify: $(6 - 5)^3 - |10 - 18|$
   - **(a)** 9
   - **(b)** 11
   - **(c)** –7
   - **(d)** –5
   - **(e)** None of these

4. Simplify: $51 - 48 \div 4(2)$
   - **(a)** 27
   - **(b)** 1.5
   - **(c)** 45
   - **(d)** $\dfrac{3}{8}$
   - **(e)** None of these

5. Simplify: $\dfrac{5(20 - 18)^2 - (-4)}{(-4)(-2)}$
   - **(a)** 2
   - **(b)** 5
   - **(c)** 3
   - **(d)** –3
   - **(e)** None of these

6. Evaluate the expression when $x = -2, y = 4, z = -3$: $x^3 + z^2 - y$
   - **(a)** –5
   - **(b)** 13
   - **(c)** –21
   - **(d)** –3
   - **(e)** None of these

7. Evaluate the expression when $x = -1, y = -2, z = 2$: $\dfrac{xz - y^2}{xy}$
   - **(a)** 3
   - **(b)** –3
   - **(c)** –1
   - **(d)** 1
   - **(e)** None of these

8. Write the statement using mathematical symbols: three times the sum of t and 12 is less than or equal to –5.
   - **(a)** $3(t + 12) < -5$
   - **(b)** $3(t + 12) \le -5$
   - **(c)** $3t + 12 \le -5$
   - **(d)** $3(t + 12) \ge -5$
   - **(e)** None of these

9. Name the property that is illustrated: $18 + x = x + 18$
   - **(a)** Commutative property for addition
   - **(b)** Associative property for multiplication
   - **(c)** Distributive property
   - **(d)** Associative property for addition
   - **(e)** None of these

10. Write an expression for the total amount of money (in cents) in d dimes and p pennies.
    - **(a)** $0.10d + 0.01p$
    - **(b)** $5d + p$
    - **(c)** $10d + p$
    - **(d)** $10d + 25p$
    - **(e)** None of these

11. Solve the equation: $9x - 32 = 5x - 4$
    - **(a)** $\{7\}$
    - **(b)** $\{-9\}$
    - **(c)** $\{9\}$
    - **(d)** $\{-7\}$
    - **(e)** None of these

12. Solve the equation: $\dfrac{h}{3} + \dfrac{3h}{6} = 10$
    - **(a)** $\{2\}$
    - **(b)** $\{300\}$
    - **(c)** $\{12\}$
    - **(d)** $\{15\}$
    - **(e)** None of these

13. Solve the equation: $|10 - 2x| = 16$
    - **(a)** $\{-13, 3\}$
    - **(b)** $\{-3\}$
    - **(c)** $\{13\}$
    - **(d)** $\{-3, 13\}$
    - **(e)** None of these

14. Solve for the specified variable: $13x - 4y = -14$; y

   (a) $y = \dfrac{-14 - 13x}{4}$ (b) $y = \dfrac{13x + 14}{4}$ (c) $y = \dfrac{13x - 14}{-4}$ (d) $y = \dfrac{4x + 14}{13}$ (e) None of these

15. Solve for the specified variable: $P = 2w + 2l$; w

   (a) $w = \dfrac{P}{2} - 2l$ (b) $w = \dfrac{P + 2l}{2}$ (c) $w = \dfrac{P - 2l}{2}$ (d) $w = \dfrac{P - l}{2}$ (e) None of these

16. Solve the inequality: $8 + \dfrac{x}{3} \geq 14$

   (a) $(2, \infty)$ (b) $(18, \infty)$ (c) $[18, \infty)$ (d) $[66, \infty)$ (e) None of these

17. Solve the inequality: $13 < 2x + 7 \leq 31$

   (a) $(3, 12]$ (b) $(10, 19)$ (c) $[3, 12)$ (d) $(10, 19]$ (e) None of these

18. Solve the inequality: $|4x - 8| > 20$

   (a) $\left(-\infty, -7\right] \cup \left[3, \infty\right)$ (b) $\left(-\infty, -3\right] \cup \left[7, \infty\right)$ (c) $\left(-\infty, -7\right) \cup \left(3, \infty\right)$ (d) $\left(-\infty, -3\right) \cup \left(7, \infty\right)$

   (e) None of these

19. Find 34% of 478. (Round to the nearest tenth.)

   (a) 162.5   (b) 1405.9   (c) 478.3   (d) 165.6   (e) None of these

20. Find the amount of money in an account after 5 years if a principal of $7,500 was invested at 5.4% interest compounded monthly and no withdrawals were made. (Round to the nearest cent.)

   (a) $175,994.02   (b) $7670.28   (c) $9818.78   (d) $48,073.01   (e) None of these

21. Name the quadrant in which the point is located: $(-2, -8)$

   (a) Quadrant I   (b) Quadrant II   (c) Quadrant III   (d) Quadrant IV   (e) None of these

22. Graph the line: $x - 2y = 8$

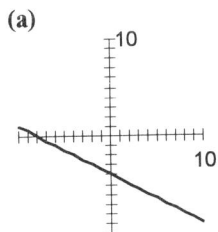

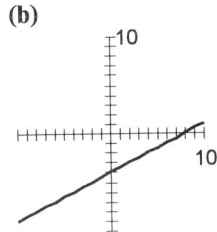

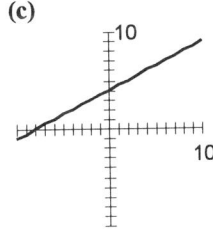

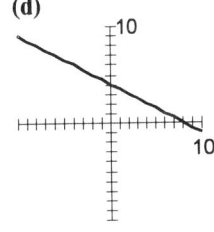

   (a)   (b)   (c)   (d)

   (e) None of these

23. Graph the line: $-4x + y = 6$

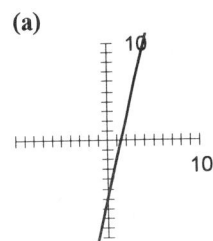

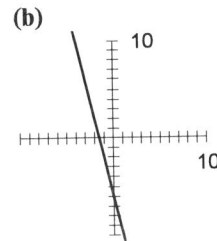

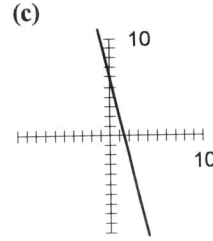

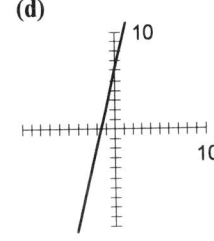

   (a)   (b)   (c)   (d)

   (e) None of these

24. Graph the line: $2x + 3y + 9 = 0$

(a)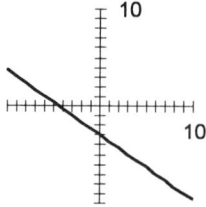

(b)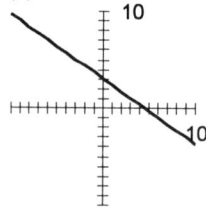

(c)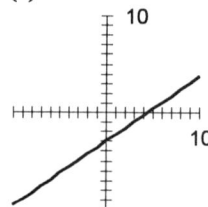

(d)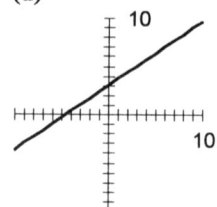

(e) None of these

25. Is the following statement true or false? The graph of $y = -3$ is a vertical line.

(a) True   (b) False   (c) Not enough information

26. Find an equation of a vertical line through (–4, 7).

(a) $y = 7$ (b) $x = 7$ (c) $x = -4$ (d) $y = -4$ (e) None of these

27. Find an equation of a line parallel to $y = -2$ through (5, –12).

(a) $y = 5$ (b) $y = -12$ (c) $x = -12$ (d) $x = 5$ (e) None of these

28. Find an equation of a line through (2, 6) with slope 3.

(a) $y = -3x + 2$ (b) $y = 3x + 6$ (c) $y = -3x$ (d) $y = 3x$ (e) None of these

29. Find an equation of a line through (–3, 5) and perpendicular to $y = \frac{3}{4}x + 3$.

(a) $y = \frac{4}{3}x + 1$ (b) $y = -\frac{3}{4}x + 3$ (c) $y = -\frac{4}{3}x + 3$ (d) $y = -\frac{4}{3}x + 1$ (e) None of these

30. Is the following relation a function? {(1, 2), (3, 5), (–2, 9), (3, –3), (4, 5)}

(a) Yes   (b) No   (c) Not enough information

31. Evaluate the determinant $\begin{vmatrix} -1 & -6 \\ 5 & 8 \end{vmatrix}$.

(a) 22   (b) –38   (c) –22   (d) 46   (e) None of these

32. Evaluate the determinant $\begin{vmatrix} 1 & 9 & 1 \\ 2 & -7 & 0 \\ 0 & 5 & -3 \end{vmatrix}$.

(a) 80   (b) –53   (c) 85   (d) 39   (e) None of these

33. Solve the system of equations: $\begin{cases} 2x - y = 11 \\ x - 5y = 10 \end{cases}$

(a) {(5, –1)}   (b) {(–1, 5)}   (c) {(0, 6)}   (d) {(11, 10)}   (e) None of these

34. Solve the system of equations: $\begin{cases} 2z = 16 \\ -2x + 3y - z = 0 \\ x + z = 13 \end{cases}$

(a) {(5, –6, 8)}   (b) {(5, 6, 8)}   (c) {(–5, 0, 8)}   (d) {(1, 2, 3)}   (e) None of these

35. If you were to solve the following system of equations with Cramer's Rule, find D_x . $\begin{cases} 8x + 2y = 4 \\ 5x - 9y = -3 \end{cases}$

(a) –82   (b) –44   (c) –30   (d) –96   (e) None of these

36. Solve the system using Cramer's Rule: $\begin{cases} 3x - 2y = 4 \\ x + 8y = -3 \end{cases}$

(a) $\left\{ \left(1, \dfrac{1}{2}\right) \right\}$ (b) $\{(2,-3)\}$ (c) $\left\{ \left(1, -\dfrac{1}{2}\right) \right\}$ (d) $\{(1, 0)\}$ (e) None of these

37. Solve the system of equations: $\begin{cases} x - 8y = 12 \\ x - 3y = -3 \end{cases}$

(a) $\{(-12, -3)\}$ (b) $\{(12, -3)\}$ (c) $\{(-4, -3)\}$ (d) $\{(-12, 3)\}$ (e) None of these

38. Solve the system of equations: $\begin{cases} 2x + y - 2z = 4 \\ y + 2z = 6 \\ -x - 2y + 3z = -6 \end{cases}$

(a) $\{1, 4, 1\}$ (b) $\{-1, 4, -1\}$ (c) $\{2, 1, 2\}$ (d) $\{0, -4, 0\}$ (e) None of these

39. A coffee shop mixes its own house blend from Premium Dark Roast Coffee and Thunder Bay Coffee. How many pounds of Premium Dark Roast Coffee (costing $8 per pound) and Thunder Bay Coffee (costing $9.80 per pound) are needed to make 40 pounds of the house blend which sells for $8.90 per pound?
(a) 10 pounds Premium Dark Roast and 20 pounds Thunder Bay Coffee
(b) 30 pounds Premium Dark Roast and 10 pounds Thunder Bay Coffee
(c) 25 pounds Premium Dark Roast and 15 pounds Thunder Bay Coffee
(d) 20 pounds Premium Dark Roast and 20 pounds Thunder Bay Coffee
(e) None of these

40. A bed & breakfast inn near Sacramento, California, charges $95 including tax per night for a room on a weekend. During the week, rooms cost $78 including tax per night. If the total receipts for one week were $3921 and a total of 47 rooms were rented, how many rooms were rented during the week and how many were rented on the weekend?
(a) 15 during the week and 32 on the weekend
(b) 45 during the week and 2 on the weekend
(c) 32 during the week and 15 on the weekend
(d) 20 during the week and 27 on the weekend
(e) None of these

41. Use positive exponents to simplify the expression: $(2x^{-2}y^3)^{-4}$

(a) $\dfrac{16x^8}{y^{12}}$ (b) $\dfrac{x^8}{16y^{12}}$ (c) $\dfrac{y^{12}}{16x^8}$ (d) $\dfrac{16y^{12}}{x^8}$ (e) None of these

42. Use positive exponents to simplify the expression: $\dfrac{3x^7y^{-5}}{12x^4y^9}$

(a) $\dfrac{1}{4x^4y^5}$ (b) $\dfrac{y^4}{4x^3}$ (c) $\dfrac{x^3y^4}{4}$ (d) $\dfrac{4x^3}{y^{-14}}$ (e) None of these

43. Write in scientific notation: 0.0000294
    (a) 2.94×10^{-4} (b) 2.94×10^5 (c) 2.94×10^4 (d) 2.94×10^{-5} (e) None of these

44. Write without exponents: 8.537×10^6
    (a) 853,700   (b) 8,537,000   (c) 8,537,000,000   (d) 0.000008537   (e) None of these

45. Perform the indicated operations: $(3x^2 - 7x + 2) - (6x^2 + 2x - 5)$
    (a) $3x^2 - 9x + 7$ (b) $-3x^2 - 9x - 3$ (c) $-3x^2 - 9x + 7$ (d) $3x^2 - 5x - 3$ (e) None of these

46. Perform the indicated operations: $(2x - 9)(2x - 9)$
    (a) $4x^2 + 36x - 81$ (b) $4x^2 - 36x + 81$ (c) $4x^2 + 81$ (d) $4x^2 - 81$ (e) None of these

47. Factor completely: $45x^4 - 75x^3 + 5x^2$
    (a) $5x^2(9x^2 - 15x + 1)$ (b) $5x(9x^3 - 15x^2 + x)$ (c) $x^2(45x^2 - 75x + 5)$ (d) None of these

48. Factor completely: $x^2 + 13x + 36$
    (a) $(x - 9)(x - 4)$ (b) $(x + 6)(x + 6)$ (c) $(x - 9)(x + 4)$ (d) $(x + 9)(x + 4)$ (e) None of these

49. Solve for the variable: $(6q + 5)(6q - 5) = 35q^2 + 11$
    (a) $\{-2, 3)$ (b) $\{-5, 7\}$ (c) $\{-6, 6\}$ (d) $\{-\sqrt{14}, \sqrt{14}\}$ (e) None of these

50. Graph: $y = -2x^2 - 5x + 7$

    (a)   (b)   (c)   (d)
    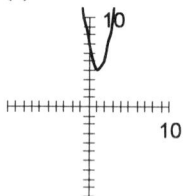   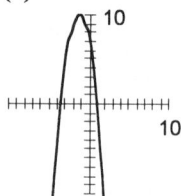   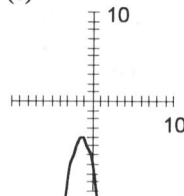   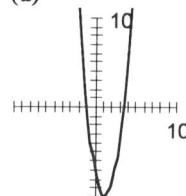

    (e) None of these

51. Find the domain of the rational function: $f(x) = \dfrac{x^2 + 7x + 6}{5x - 20}$
    (a) $\{x | x \text{ is a real number and } x \neq -1, -6\}$ (b) $\{x | x \text{ is a real number and } x \neq -4\}$
    (c) $\{x | x \text{ is a real number and } x \neq 4\}$ (d) $\{x | x \text{ is a real number and } x \neq 1, 6\}$ (e) None of these

52. Write the rational expression in lowest terms: $\dfrac{x^2 + 5x + 4}{x^2 + 7x + 12}$
    (a) $\dfrac{x^2 + 5x + 4}{x^2 + 7x + 12}$ (b) $\dfrac{x + 1}{x + 3}$ (c) $\dfrac{1}{3}$ (d) $\dfrac{x + 1}{x - 3}$ (e) None of these

53. Perform the indicated operation and reduce to lowest terms: $\dfrac{y^3}{y^2 - 9} \cdot \dfrac{y - 3}{2y^3}$
    (a) $\dfrac{1}{2(y + 3)}$ (b) $\dfrac{1}{2(y - 3)}$ (c) $\dfrac{y^4 - 3y^3}{2y^5 - 18y^3}$ (d) $\dfrac{y}{2(y - 3)}$ (e) None of these

54. Perform the indicated operation and reduce to lowest terms : $\dfrac{-6x-3}{x^2+12x+11}+\dfrac{2}{x+1}$

(a) $\dfrac{-6x-1}{x^2+12x+11}$ (b) $\dfrac{-4x+19}{x^2+12x+11}$ (c) $\dfrac{-5x-14}{x+11}$ (d) $\dfrac{4x-19}{x^2+12x+11}$ (e) None of these

55. Divide and simplify your answer: $\dfrac{\dfrac{5}{2}+\dfrac{9}{3}}{1-\dfrac{1}{2}}$

(a) 7   (b) –65   (c) 11   (d) 32   (e) None of these

56. Divide and simplify your answer: $\dfrac{3a^2b^4c^3-27a^{-1}bc+45ac^2}{9abc}$

(a) $\dfrac{ab^3c^2}{3}-\dfrac{1}{3a^2}+\dfrac{c}{5b}$ (b) $-\dfrac{ab^3c^2}{3a}-\dfrac{3}{a^2}+\dfrac{5c^2}{b}$ (c) $\dfrac{ab^5c^2}{3}-\dfrac{3}{a^{-2}}+\dfrac{5c}{b}$

(d) $\dfrac{ab^3c^2}{3}-\dfrac{3}{a^2}+\dfrac{5c}{b}$ (e) None of these

57. Use synthetic division to divide $2x^3-9x^2+14x-13$ by $x-4$

(a) $2x^3-x^2+72x+275$ (b) $-2x^3-x^2+10x+27$ (c) $2x^2-x+72+\dfrac{275}{x-4}$

(d) $2x^2-x+10+\dfrac{27}{x-4}$ (e) None of these

58. Solve for x: $\dfrac{x-5}{2x-7}=-1$

(a) $\{-4\}$ (b) $\{4\}$ (c) $\left\{-\dfrac{2}{3}\right\}$ (d) $\{12\}$ (e) None of these

59. Suppose that R is directly proportional to w. If $R=604$ when $w=34$, find R when $w=102$.

(a) $\dfrac{604}{3}$ (b) 5.74   (c) 1812   (d) 928   (e) None of these

60. Suppose that t is indirectly proportional to the square of x. If $t=15$ when $x=2$, find t when $x=5$.
(a) 2.4   (b) 12   (c) 1.5   (d) 6   (e) None of these

61. Simplify. Assume that all variables represent positive numbers. Write using only positive exponents:
$-\sqrt[5]{x^{32}y^{15}}$

(a) $-x^{27}y^{10}$ (b) $-x^5y^{10}\sqrt[5]{x^2}$ (c) $x^6y^4\sqrt[5]{x^2}$ (d) $-x^6y^3\sqrt[5]{x^2}$ (e) None of these

62. Simplify. Write using only positive exponents: $\left(\dfrac{1}{4913}\right)^{-1/2}$

(a) $\dfrac{1}{70}$ (b) $17^{3/2}$ (c) $\dfrac{1}{17^{3/2}}$ (d) 70   (e) None of these

63. Rationalize the denominator: $\sqrt{\dfrac{35}{3}}$

  (a) $\dfrac{\sqrt{35}}{3}$ (b) $\dfrac{\sqrt{105}}{3}$ (c) $\dfrac{35}{\sqrt{105}}$ (d) $\dfrac{105}{9}$ (e) None of these

64. Perform the indicated operations and simplify. Assume that variables are positive: $\sqrt{64q^3} - \sqrt{9q}$

  (a) $5\sqrt{q}$ (b) $8q\sqrt{q} - 3\sqrt{q}$ (c) $5q\sqrt{q}$ (d) $8\sqrt{q^3} - 3\sqrt{q}$ (e) None of these

65. Perform the indicated operations: $\left(\sqrt{3} + 9\right)\left(\sqrt{3} - 9\right)$

  (a) -78 (b) -72 (c) 84 (d) 90 (e) None of these

66. Solve the equation for x: $x = \sqrt{-x + 20}$

  (a) $\{-5, 4\}$ (b) $\{-4, 5\}$ (c) $\{4\}$ (d) $\{2\}$ (e) None of these

67. Perform the indicated operation, simplify, and write the result in the form $a + bi$: $(-4 + 3i) - (10 - 2i)$

  (a) $-14 + i$ (b) $6 + 5i$ (c) -19 (d) $-14 + 5i$ (e) None of these

68. Perform the indicated operation, simplify, and write the result in the form $a + bi$: $(4 + 9i)(4 - 9i)$

  (a) $16 - 81i$ (b) $4 - 81i$ (c) 97 (d) -65 (e) None of these

69. Solve for the variable: $(x - 3)^2 = 40$

  (a) $\left\{3 - 2\sqrt{5}, \, 3 + 2\sqrt{5}\right\}$ (b) $\left\{3 - 2\sqrt{10}, \, 3 + 2\sqrt{10}\right\}$ (c) $\left\{-3 - 2\sqrt{10}, \, -3 + 2\sqrt{10}\right\}$ (d) $\{5, 8\}$
  (e) None of these

70. Solve for the variable: $y^2 + 10y - 9 = 0$

  (a) $\left\{-5 - \sqrt{34}, \, -5 + \sqrt{34}\right\}$ (b) $\{-9, -1\}$ (c) $\{1, 9\}$ (d) $\left\{5 - \sqrt{34}, \, 5 + \sqrt{34}\right\}$ (e) None of these

71. Solve for the variable: $b^2 - 13b = -22$

  (a) $\left\{\dfrac{131 - \sqrt{257}}{2}, \, \dfrac{131 + \sqrt{257}}{2}\right\}$ (b) $\{-2, -11\}$ (c) $\{0, 7\}$ (d) $\{2, 11\}$ (e) None of these

72. Solve for the variable: $a^4 - 13a^2 + 36 = 0$

  (a) $\{-3i, -2i, 2i, 3i\}$ (b) $\{-3, -2, 2, 3\}$ (c) $\{4, 9\}$ (d) $\{-4, -9\}$ (e) None of these

73. Solve for the variable: $(x - 5)^2 - 10(x - 5) - 39 = 0$

  (a) $\{-8, 8\}$ (b) $\{-3, 13\}$ (c) $\{2, 18\}$ (d) $\{-13, 3\}$ (e) None of these

74. Solve the inequality for x, and write the solution set in interval notation: $x^2 + 5x \ge -6$

  (a) $[-3, -2]$ (b) $(-\infty, -3] \cup [-2, \infty)$ (c) $(-3, -2)$ (d) $(-\infty, -3) \cup (-2, \infty)$ (e) None of these

75. Graph the function: $h(x) = -x^2 + 2x - 3$

(a)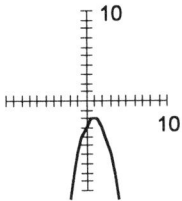

(b)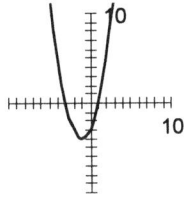

(c)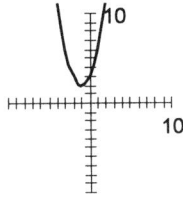

(d)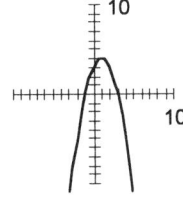

(e) None of these

76. A 16-foot ladder is leaning against a house. The distance from the bottom of the ladder to the house is 9 feet less than the distance from the top of the ladder to the ground. Find how far the top of the ladder is from the ground. Give a one-decimal place approximation.

(a) 5.9 feet   **(b)** 18.2 feet   **(c)** 14.9 feet   **(d)** 9.2 feet   **(e)** None of these

77. Find the distance between the points: $(-3, 12), (8, 5)$

(a) $\sqrt{170}$ **(b)** $\sqrt{74}$ **(c)** $\sqrt{314}$ **(d)** $\sqrt{410}$ **(e)** None of these

78. Find the midpoint of the segment joining the points $(1, -5), (7, -9)$

(a) $(0, 4)$ **(b)** $(-3, 2)$ **(c)** $(3, -2)$ **(d)** $(4, -7)$ **(e)** None of these

79. Identify the graph of the equation: $2x^2 + 2y^2 = 16$

(a) parabola   **(b)** circle   **(c)** ellipse   **(d)** hyperbola   **(e)** None of these

80. Identify the graph of the equation: $\dfrac{y^2}{36} = 1 + \dfrac{x^2}{15}$

(a) parabola   **(b)** circle   **(c)** ellipse   **(d)** hyperbola   **(e)** None of these

81. Identify the graph of the equation: $2x^2 + 3y^2 = 14$

(a) parabola   **(b)** circle   **(c)** ellipse   **(d)** hyperbola   **(e)** None of these

82. Identify the graph of the equation: $x - (y + 8)^2 = 12$

(a) parabola   **(b)** circle   **(c)** ellipse   **(d)** hyperbola   **(e)** None of these

83. Solve the system: $\begin{cases} x = 2(y-3)^2 + 5 \\ y = 4 \end{cases}$

(a) $\{(5, 3)\}$ **(b)** $\{(0, 5)\}$ **(c)** $\{(7, 4)\}$ **(d)** $\{(4, -7)\}$ **(e)** None of these

84. If $f(x) = 3x - 1$, and $g(x) = 2x^2$, find $(g \circ f)(x)$.

(a) $2x^2 + 3x - 1$ **(b)** $6x^3 - 1$ **(c)** $18x^2 - 12x + 2$ **(d)** $6x^2 - 1$ **(e)** None of these

85. If $f(x) = 4x^4 - x^3 + 6x^2 + 15x - 1$, and $g(x) = 3x^2 + 4$, find $(f \circ g)(0)$.

(a) 7   **(b)** 1115   **(c)** 93   **(d)** 79   **(e)** None of these

86. Find the inverse of the one-to-one function: $f(x) = 8x^3 + 2$

(a) $\dfrac{\sqrt[3]{x-2}}{2}$ **(b)** $-\dfrac{1}{4x^3}$ **(c)** $\dfrac{\sqrt[3]{x+2}}{2}$ **(d)** $8\sqrt[3]{x} - 2$ **(e)** None of these

87. Is the function one-to-one? $f = \{(9, -5), (-1, 5), (1, -2), (4, 2)\}$
    (a) Yes   **(b)** No   **(c)** Not enough information

88. Use the properties of logarithms to write the expression as a single logarithm: $3\log_9 2 - 2\log_9 x + 4\log_9 y$
    (a) $\dfrac{8y^4}{x^2}$ **(b)** $\log_9 \dfrac{8y^4}{x^2}$ **(c)** $\log_9 \dfrac{16y}{x}$ **(d)** $\log_9 8x^2 y^4$ **(e)** None of these

89. Write the expression $\log_2 \dfrac{15q^5}{r}$ as the sum or difference of multiples of logarithms.
    (a) $\log_2 15 - \log_2 (q^5 - r)$ **(b)** $\log_2 15 - 5\log_2 q + \log_2 r$ **(c)** $\log_2 15 + 5\log_2 (q - r)$
    (d) $\log_2 15 + 5\log_2 q - \log_2 r$ **(e)** None of these

90. Solve $2^{7-3x} = 512$ for x. Give an exact solution.
    (a) $\{5\}$ **(b)** $\left\{-\dfrac{2}{3}\right\}$ **(c)** $\{-2\}$ **(d)** $\left\{-\dfrac{3}{2}\right\}$ **(e)** None of these

91. Solve the logarithmic equation for x: $\log_3 27 + \log_3 x = 3$. Give an exact solution.
    (a) $\{10\}$ **(b)** $\{1\}$ **(c)** $\{1, 0\}$ **(d)** $\{8\}$ **(e)** None of these

C 92. Use a grapher to approximate the solution of $e^{x+7} = 9 - x$ to two decimal places.
    (a) $\{-9.97\}$ **(b)** $\{1.51\}$ **(c)** $\{-4.40\}$ **(d)** $\{5.74\}$ **(e)** None of these

93. Find the third term of the sequence $a_n = \dfrac{3n^2}{10 - n^2}$.
    (a) 27   **(b)** $\dfrac{1}{3}$ **(c)** -8 **(d)** 2   **(e)** None of these

94. Find the fortieth term of the sequence $a_n = 50 - 8(n - 1)$.
    (a) 262   **(b)** -262 **(c)** -54 **(d)** -270 **(e)** None of these

95. Find the general term of the sequence $\dfrac{1}{2}, -\dfrac{1}{4}, \dfrac{1}{8}, \ldots$
    (a) $a_n = \dfrac{1}{2} + \left(\dfrac{1}{2}\right)^n$ **(b)** $a_n = -\dfrac{1}{2}\left(\dfrac{1}{2}\right)^{n-1}$ **(c)** $a_n = \dfrac{1}{2}\left(-\dfrac{1}{2}\right)^{n-1}$ **(d)** None of these

96. Find the partial sum S_3 of the sequence $a_n = 2(3)^{2n+1}$.
    (a) 26   **(b)** -1 **(c)** 78   **(d)** 4914   **(e)** None of these

97. Find the infinite sum S_∞ of the sequence $a_1 = 41$ and $r = \dfrac{2}{3}$.
    (a) 123   **(b)** 61.5   **(c)** 138   **(d)** 82   **(e)** None of these

98. Find the sum: $\displaystyle\sum_{i=4}^{7} (5i - 5)$
    (a) 0   **(b)** 90   **(c)** 105   **(d)** 51   **(e)** None of these

99. Is the following statement true or false? The fourth term of the expansion of $(w - z)^7$ is $-35w^4z^3$.

   (a) True   **(b)** False   **(c)** Not enough information

100. A gardener is planting shrubs to fill a plot of land in the shape of a trapezoid with 3 shrubs in the first row, 7 shrubs in the second row, 11 shrubs in the third row, and so on for 9 rows. How many shrubs will the gardener need?

   (a) 28   **(b)** 105   **(c)** 171   **(d)** 697   **(e)** None of these

Answers to Exercises

CHAPTER 1

Section 1.1

1. $\frac{1}{2}x = \frac{1}{2}(4) = 2$
2. 15
3. $-14z = -14(2) = -28$
4. -3
5. $xy = (2)(12) = 24$
6. 1
7. $2ab - 3 = 2(4)(5) - 3$
    $= 40 - 3 = 37$
8. 0
9. $bh = (15)(32)$
    $= 480$ square inches
10. 4.25 square centimeters
11. $\{0, 1, 2\}$
12. $\varnothing$
13. $\{1, 2, 3, 4, 5\}$
14. $\{2, 4, 6, 8, \dots\}$
15. $\{2, \sqrt{49}, 0, \frac{12}{3}\}$
16. $\{2, \frac{17}{4}, -\frac{1}{2}, \sqrt{49}, 0, -\sqrt{49}, \frac{12}{3}\}$
17. $\{2, \frac{17}{4}, \pi, -\frac{1}{2}, \sqrt{49}, 0, -\sqrt{49}, \frac{12}{3}, \sqrt{50}\}$
18. $\{2, \sqrt{49}, 0, -\sqrt{49}, \frac{12}{3}\}$
19. $\{2, \sqrt{49}, \frac{12}{3}\}$
20. $\{\pi, \sqrt{50}\}$
21. $2z$
22. $\frac{1}{2}z - 3$
23. $\frac{z}{11}$
24. $z - 10$
25. $z - 4z$
26. $100z + 6$
27. $9f \div c = 9(6) \div 360 = 54 \div 360 = 0.15$ or 15%
28. 977,250 gallons
29. (a) Brunswick
    (b) Polaris Industries
    (c)
        $2,836,000,000
        $- \$862,000,000$
        $2,010,000,000

Section 1.2

1. $\dfrac{-7}{y} = 6$
2. $9(5 - z) = -3$
3. $9 + z = 18$
4. $3(b + 7) = 56$
5. $\dfrac{t}{3} < 11$
6. $4 - z \neq 15$
7. $2(q + 8) \geq -12$
8. $5(z - 13) \leq 2$
9. (a) $\dfrac{3}{13}$
    (b) $-\dfrac{13}{3}$
10. (a) 0
    (b) undefined
11. (a) 4
    (b) $-\dfrac{1}{4}$
12. (a) -17
    (b) $\dfrac{1}{17}$
13. (a) $-\dfrac{5}{3}$
    (b) $\dfrac{3}{5}$
14. (a) $-\dfrac{1}{9}$
    (b) 9
15. (a) $\dfrac{21}{2}$
    (b) $-\dfrac{2}{21}$
16. (a) 9
    (b) $-\dfrac{1}{9}$
17. Commutative Property of Addition
18. Associative Property of Addition
19. Associative Property of Multiplication
20. Commutative Property of Multiplication
21. Distributive Property
22. Associative Property of Addition
23. Distributive Property
24. $x(-3)$
25. $p + 4q = 4q + p$

26. zy

27. $3r + s = s + 3r$

28. $5r + (s + 3t)$

29. $2w + (3v + 17u) = (2w + 3v) + 17u$

30. $(3a)b$

31. $(-5.3x)y = -5.3(xy)$

32. $2 + 2r$

33. $z(7 + 6w) = z(7) + z(6w) = 7z + 6wz$

34. $3x + 3y + 3z$

35. $7(2a + 2b + 2c) = 7(2a) + 7(2b) + 7(2c)$
$$= 14a + 14b + 14c$$

Section 1.3

1. $|-16| = 16$

2. 55

3. $-|-2| = -(2) = -2$

4. -7

5. $-4 + (-3) = -7$

6. -10

7. $5 - 12 = 5 + (-12) = -7$

8. 25

9. $-4(12) = -48$

10. 30

11. $-14 \div -7 = \dfrac{-14}{-7} = 2$

12. -6

13. $(-4)^2 = (-4)(-4) = 16$

14. -16

15. $5^3 = 5 \cdot 5 \cdot 5 = 125$

16. $\dfrac{1}{27}$

17. $\sqrt{121} = 11$ because $11^2 = 121$

18. 5

19. $\sqrt[4]{81} = 3$ because $3^4 = 81$

20. $\dfrac{2}{5}$

21. $65 - (-27) = 65 + 27 = 92$ dollars

22. $40

23. $36 - (-7) = 36 + 7 = 43°$ Fahrenheit

24. $48.6°$ Fahrenheit

Section 1.4

1. $3 \cdot 2^2 - 5 = 3 \cdot 4 - 5 = 12 - 5 = 7$

2. -17

3. $\dfrac{15 + 3(-4)}{\sqrt{9}} = \dfrac{15 + (-12)}{\sqrt{9}} = \dfrac{15 - 12}{3} = \dfrac{3}{3} = 1$

4. 2

5. $3x - y + 2z = 3(2) - (10) + 2(-1)$
$$= 6 - 10 - 2 = -6$$

6. 10

7. $-10(x + y + 5z) = -10(2 + 10 + 5(-1))$
$$= -10(2 + 10 - 5)$$
$$= -10(7) = -70$$

8. 5

9. (a) $\dfrac{24c}{a(p+1)} = \dfrac{24(100)}{1000(23+1)}$
$$= \dfrac{2400}{1000(24)}$$
$$= \dfrac{2400}{24,000} = 0.1$$
or 10%

 (b) approximately 2%

10. (a) $2lw + 2hl + 2hw$
$$= 2(80)(40) + 2(30)(80) + 2(30)(40)$$
$$= 6400 + 4800 + 2400$$
$$= 13,600 \text{ square centimeters}$$

 (b) 10,400 square inches

11.

x	-1	0	1	2
$2x^2 - 3$	-1	-3	-1	5

$x = -1$: $2(-1)^2 - 3 = 2(1) - 3 = 2 - 3 = -1$

$x = 0$: $2(0)^2 - 3 = 2(0) - 3 = 0 - 3 = -3$

$x = 1$: $2(1)^2 - 3 = 2(1) - 3 = 2 - 3 = -1$

$x = 2$: $2(2)^2 - 3 = 2(4) - 3 = 8 - 3 = 5$

12.

y	-3	-1	2	4
$-5(4y + 10)$	10	-30	-90	-130

13. $1.65x$

14. $1.179y$

15. $0.45x + 0.39y$

16. $3.49c + 0.49d$

17. $80 - z$

18. $243 - q$

19. $5.5 - x$

20. $y - 10$

21. $2x - 5y + 4 - x + 10y = 2x - x - 5y + 10y + 4$
$$= x + 5y + 4$$

22. $8a + 27b$

23. $2r(3 + r) = 2r(3) + 2r(r) = 6r + 2r^2$

24. $-3t + 5$

25. $-3(2u - 5) - 2(10u + 1)$
$$= -3(2u) + (-3)(-5) + (-2)(10u) + (-2)(1)$$
$$= -6u + 15 - 20u - 2$$
$$= -26u + 13$$

26. $13n - 5$
27. $4x - 7y$
28. $-10b - 17$
29. 3,600,000 cars
30. 3,900,000 cars
31. 1989
32. 1992
33. $3,950,000 - 3,450,000 = 500,000$ cars
34. 0 cars

Chapter 1 Practice Test

1. True
2. True
3. True
4. False
5. False
6. False
7. 5
8. 18
9. −52
10. −3
11. 2
12. 5
13. $\dfrac{2}{3}$
14. (a) 24.95, 99.80, 249.50, 748.50
   (b) Increase
15. $3\left|12 - z\right| = 24$
16. $\dfrac{(x+7)^3}{5} \geq 0$
17. $2a + 9 \neq -a$
18. $-5 < \dfrac{6y}{\left|y+2\right|}$
19. Associative Property of Addition
20. Multiplicative Inverse Property
21. Commutative Property of Multiplication
22. Distributive Property
23. $p + 25\,q$

Chapter 2

Section 2.1

1.
$$12 - 2y = 6$$
$$12 - 2y - 12 = 6 - 12$$
$$-2y = -6$$
$$\frac{-2y}{-2} = \frac{-6}{-2}$$
$$y = 3$$
The solution set is $\{3\}$.

2. $\{15\}$
3.
$$8x + 8 = -16$$
$$8x + 8 - 8 = -16 - 8$$
$$8x = -24$$
$$\frac{8x}{8} = \frac{-24}{8}$$
$$x = -3$$
The solution set is $\{-3\}$.

4. $\{10.56\}$
5.
$$55 = 4 + 17w$$
$$55 - 4 = 4 + 17w - 4$$
$$51 = 17w$$
$$\frac{51}{17} = \frac{17w}{17}$$
$$3 = w$$
The solution set is $\{3\}$.

6. $\{13.25\}$
7.
$$-30 = -5q + 20$$
$$-30 - 20 = -5q + 20 - 20$$
$$-50 = -5q$$
$$\frac{-50}{-5} = \frac{-5q}{-5}$$
$$10 = q$$
The solution set is $\{10\}$.

8. $\{6\}$
9.
$$10t - 3 = 5t + 12$$
$$10t - 3 - 5t = 5t + 12 - 5t$$
$$5t - 3 = 12$$
$$5t - 3 + 3 = 12 + 3$$
$$5t = 15$$
$$\frac{5t}{5} = \frac{15}{5}$$
$$t = 3$$
The solution set is $\{3\}$.

10. $\left\{\dfrac{29}{8}\right\}$
11.
$$17 - 8y = -5y - 13$$
$$17 - 8y + 8y = -5y - 13 + 8y$$
$$17 = 3y - 13$$
$$17 + 13 = 3y - 13 + 13$$
$$30 = 3y$$
$$\frac{30}{3} = \frac{3y}{3}$$
$$10 = y$$
The solution set is $\{10\}$.

12. $\left\{\dfrac{20}{11}\right\}$

13.
$$-2(x-6) = x+15$$
$$-2x+12 = x+15$$
$$-2x+12-x = x+15-x$$
$$-3x+12 = 15$$
$$-3x+12-12 = 15-12$$
$$-3x = 3$$
$$\dfrac{-3x}{-3} = \dfrac{3}{-3}$$
$$x = -1$$
The solution set is $\{-1\}$.

14. $\left\{-\dfrac{29}{8}\right\}$

15.
$$3(5-x) = 12-x$$
$$15-3x = 12-x$$
$$15-3x+x = 12-x+x$$
$$15-2x = 12$$
$$15-2x-15 = 12-15$$
$$-2x = -3$$
$$\dfrac{-2x}{-2} = \dfrac{-3}{-2}$$
$$x = \dfrac{3}{2}$$
Solution

16. Solution

17.
$$12(x-3) = 4(3x-9)$$
$$12x-36 = 12x-9$$
$$12x-36-12x = 12x-9-12x$$
$$-36 = -9$$
No solution

18. Identity

19.
$$9x-4 = 5(x+3)+4x$$
$$9x-4 = 5x+15+4x$$
$$9x-4 = 9x+15$$
$$9x-4-9x = 9x+15-9x$$
$$-4 = 15$$
No solution

20. Identity

21.
$$\dfrac{1+2x}{3} = \dfrac{3-4x}{5}$$
$$15\left(\dfrac{1+2x}{3}\right) = 15\left(\dfrac{3-4x}{5}\right)$$
$$5+10x = 9-12x$$
$$5+10x+12x = 9-12x+12x$$
$$5+22x = 9$$
$$5+22x-5 = 9-5$$
$$22x = 4$$
$$\dfrac{22x}{22} = \dfrac{4}{22}$$
$$x = \dfrac{2}{11}$$
The solution set is $\left\{\dfrac{2}{11}\right\}$.

22. $\left\{\dfrac{46}{7}\right\}$

23.
$$\dfrac{2y}{3}+\dfrac{y}{6} = 15$$
$$6\left(\dfrac{2y}{3}+\dfrac{y}{6}\right) = 6(15)$$
$$4y+y = 90$$
$$5y = 90$$
$$\dfrac{5y}{5} = \dfrac{90}{5}$$
$$y = 18$$
The solution set is $\{18\}$.

24. $\left\{-\dfrac{81}{2}\right\}$

25. $y+y+1+y+2 = 3y+3$

26. $9q+3$

27. $10t+2t+14 = 12t+14$

28. $105x+300$

Section 2.2

1.
$$x-0.1x = 14{,}395.50$$
$$0.9x = 14{,}395.50$$
$$x = 15{,}995$$
The regular price of the car is $15,995.

2. $23

3. $80{,}000+80{,}000(0.15) = 92{,}000$
The house is worth $92,000 today.

4. $2450

5. $19{,}930{,}447(0.431) \approx 8{,}590{,}022.66$
The schools received about $8,590,022.66 from the state of Texas in 1994-1995.

6. $467,351.04

7.
$$x + 1.707x = 471$$
$$2.707x = 471$$
$$x \approx 174$$
The average size of a farm in the United States in 1940 was approximately 174 acres.

8. About 108,051,613 workers

9.
$$x + x + 1 + x + 2 = 141$$
$$3x + 3 = 141$$
$$3x = 138$$
$$x = 46$$
The integers are 46, 47, and 48.

10. 15, 16, and 17

11.
$$x + 2(x + 1) + x + 2 = 7860$$
$$x + 2x + 2 + x + 2 = 7860$$
$$4x + 4 = 7860$$
$$4x = 7856$$
$$x = 1964$$
The years are 1964, 1965, and 1966.

12. 0.75

13.
$$x + x + x + 15 = 180$$
$$3x + 15 = 180$$
$$3x = 165$$
$$x = 55$$
The angles of the triangle are 55°, 55°, and 70°.

14. 36°

15. (a) $100\% - 45\% - 12\% - 16\% = 27\%$
    (b) Arts and sciences students
    (c) Faculty
    (d) $300(0.16) = 48$ tickets
    (e) $300(0.27) = 81$ tickets

Section 2.3

1.
$$D = rt$$
$$\frac{D}{t} = \frac{rt}{t}$$
$$\frac{D}{t} = r$$

2.
$$y = \frac{10x + 27}{3}$$

3.
$$A = 3M - 2N$$
$$A + 2N = 3M$$
$$\frac{A + 2N}{3} = \frac{3M}{3}$$
$$\frac{A + 2N}{3} = M$$

4.
$$R = \frac{E}{I} - r$$

5.
$$A = 5H(b + B)$$
$$\frac{A}{5(b + B)} = H$$

6.
$$s = \frac{N}{3t^4 - 5v}$$

7.
$$r = \frac{d}{2} = \frac{12}{2} = 6$$
$$C = 2\pi r$$
$$C = 2\pi(6)$$
$$C = 12\pi \approx 37.7$$
The circumference of the pizza is approximately 37.7 inches.

8. Approximately 15.7 feet

9.
$$A = P\left(1 + \frac{r}{n}\right)^{nt}$$
$$A = 2000\left(1 + \frac{0.04}{4}\right)^{4 \cdot 2}$$
$$A = 2000(1.01)^8$$
$$A \approx 2165.71$$
After 2 years, there will be approximately $2165.71 in the account.

10. Approximately $1072.29

11.
$$A = P\left(1 + \frac{r}{n}\right)^{nt}$$
$$A = 5000\left(1 + \frac{0.035}{26}\right)^{26}$$
$$A = 5000(1.035595334)$$
$$A \approx 5177.98$$
After 1 year, there will be approximately $5177.98 in the account.

12. Approximately $8180.51

13. (a) Area of quilt = $l \times w = 90 \times 66 = 5940$ square inches
    Area of square = $l \times w = 6 \times 6 = 36$ square inches
    Area of quilt ÷ area of square = $5940 \div 36 =$ 165 squares
    (b) $90 \div 6 = 15$ rows
    (c) $66 \div 6 = 11$ patches per row

14. (a) 1728 tiles
    (b) 12 packages of tiles

15. $r \times t = d$
$$62t = 170$$
$$t = \frac{170}{62} \approx 2.74$$
Yes, you can reach Cleveland in less than 3 hours.

16. Approximately 57.7 miles per hour

17.
$$\text{Distance in miles} = \frac{4260}{5280}$$

$$r \times t = d$$

$$45t = \frac{4260}{5280}$$

$$t \approx 0.01793$$

It will take about 0.01793 hours or 1.08 minutes to drive over the bridge.

18. Approximately 0.353 hours or 21.2 minutes

19.
(a) P(green) = $\frac{1}{3}$

(b) P(yellow) = $\frac{1}{3}$

(c) P(not yellow) = P(red or green) = $\frac{2}{3}$

20.
(a) $\frac{1}{10}$

(b) $\frac{1}{2}$

(c) $\frac{3}{10}$

Section 2.4

1. $(28, \infty)$
2. $(-\infty, -5)$
3. $(-\infty, 16]$
4. $[-1, \infty)$
5. $[-3, 11)$
6. $[9, 17]$
7. $(-4, -2]$
8. $(-18, 8)$
9. $6 + 3x \leq 3$

$$3x \leq -3$$
$$x \leq -1$$
$$(-\infty, -1]$$

10. $(-\infty, 15)$
11. $6 - 2x \geq 2$

$$-2x \geq -4$$
$$\frac{-2x}{-2} \leq \frac{-4}{-2}$$
$$x \leq 2$$
$$(-\infty, 2]$$

12. $[3, \infty)$
13. $5(x + 2) > 2(2x - 1)$

$$5x + 10 > 4x - 2$$
$$x + 10 > -2$$
$$x > -12$$
$$(-12, \infty)$$

14. $[4, \infty)$

15. $\frac{5x + 15}{4} < 0$

$$5x + 15 < 0$$
$$5x < -15$$
$$x < -3$$
$$(-\infty, -3)$$

16. $[-6, \infty)$

17. $\frac{1}{2}(2x - 5) > 2 - x$

$$x - \frac{5}{2} > 2 - x$$
$$2x - \frac{5}{2} > 2$$
$$2x > \frac{9}{2}$$
$$x > \frac{9}{4}$$
$$\left(\frac{9}{4}, \infty\right)$$

18. 82.5

19. $\frac{65 + 82 + 93 + 3x}{6} \geq 75$

$$65 + 82 + 93 + 3x \geq 450$$
$$240 + 3x \geq 450$$
$$3x \geq 210$$
$$x \geq 70$$

The student must get a minimum score of 70 to advance to the next course.

20. (a) 2.5 pounds
    (b) 5 pounds

Section 2.5

1. $\{x | x \text{ is an integer}\}$
2. $\varnothing$
3. $\{-3, -2, -1, 0, 1, 2, 3\}$
4. $\{0\}$
5. $\{0\}$
6. $\{0, 1, 2, 3\}$
7. $\varnothing$
8. $\{-3, -2, -1\}$
9. $6 + 3x \geq 3 \quad$ and $\quad x - 6 < 0$

$$3x \geq -3$$
$$x \geq -3 \qquad \qquad x < 6$$
$$[-3, 6)$$

10. $\left(-\frac{2}{3}, 7\right)$

11. $9 \le x - 25 \le 14$

$34 \le x \le 39$

$[34, 39]$

12. $\left(-\dfrac{3}{2}, 3\right)$

13. $-12 \le 4(2 - x) < 17$

$-12 \le 8 - 4x < 17$

$-20 \le -4x < 9$

$\dfrac{-20}{-4} \ge \dfrac{-4x}{-4} > \dfrac{9}{-4}$

$5 \ge x > -\dfrac{9}{4}$

$\left(-\dfrac{9}{4}, 5\right]$

14. $\left[-\dfrac{65}{6}, -\dfrac{27}{6}\right)$

15. $x \le 0 \quad$ or $\quad 7x \ge 49$

$x \le 0 \qquad x \ge 7$

$(-\infty, 0] \cup [7, \infty)$

16. $(-\infty, -11) \cup (-2, \infty)$

17. $4x < 2 \quad$ or $\quad x - 5 \ge 12$

$x < 2 \qquad x \ge 17$

$(-\infty, 2) \cup [17, \infty)$

18. $(-\infty, 12] \cup (15, \infty)$

19. $-2.7y < 4 \qquad$ or $\quad 5.6y < 30$

$\dfrac{-2.7y}{-2.7} > \dfrac{4}{-2.7}$

$y > -1.481 \qquad\qquad y < \dfrac{30}{5.6}$

$(-\infty, \infty)$

20. $(-\infty, \infty)$

21. $26 \le \dfrac{9}{5}C + 32 \le 96$

$-6 \le \dfrac{9}{5}C \le 64$

$-\dfrac{30}{9} \le C \le \dfrac{320}{9}$

These are approximately –3.3°C to 35.6°C.

22. Approximately –31.7°C to 39.4°C

23. $82 \le \dfrac{94 + 69 + 77 + 2x}{5} < 86$

$410 \le 94 + 69 + 77 + 2x < 430$

$410 \le 240 + 2x < 430$

$170 \le 2x < 190$

$85 \le x < 95$

The student must score in the range of [85, 95) on the final exam to receive a B in the course.

24. Assuming the maximum possible score on the final exam is 100, it is impossible for the student to receive an A– in the course.

25. $200 \le 14 + 8.5x \le 350$

$186 \le 8.5x \le 336$

$21.88235294 \le x \le 39.52941176$

For the given range, your company could purchase between 22 and 39 reams of paper.

26. Between 118 and 823 T-shirts

Section 2.6

1. $|w| = 27$

$w = 27 \quad$ or $\quad w = -27$

$\{-27, 27\}$

2. $\{-6.5, 6.5\}$

3. $|5x| = 45$

$5x = 45 \quad$ or $\quad 5x = -45$

$x = 9 \quad$ or $\quad x = -9$

$\{-9, 9\}$

4. $\left\{-\dfrac{7}{2}, \dfrac{7}{2}\right\}$

5. $|10x - 6| = 54$

$10x - 6 = 54 \quad$ or $\quad 10x - 6 = -54$

$10x = 60 \quad$ or $\quad 10x = -48$

$x = 6 \quad$ or $\quad x = -4.8$

$\{-4.8, 6\}$

6. $\{-8, 15\}$

7. $|4 - 2x| = 12$

$4 - 2x = 12 \quad$ or $\quad 4 - 2x = -12$

$-2x = 8 \quad$ or $\quad -2x = -16$

$x = -4 \quad$ or $\quad x = 8$

$\{-4, 8\}$

8. $\left\{-8, \dfrac{8}{3}\right\}$

9. $|x| - 7 = -3$

$|x| = 4$

$x = 4 \quad$ or $\quad x = -4$

$\{-4, 4\}$

10. $\{-69, 69\}$

11. $|2x| - 4 = 35$

$|2x| = 39$

$2x = 39 \quad$ or $\quad 2x = -39$

$x = 19.5 \quad$ or $\quad x = -19.5$

$\{-19.5, 19.5\}$

12. $\{-12, 12\}$

13. $|x + 2| = |2x - 5|$

$x + 2 = 2x - 5$ or $x + 2 = -(2x - 5)$
$\qquad\qquad\qquad\qquad x + 2 = -2x + 5$

$\quad 2 = x - 5$ or $3x = 3$

$\quad 7 = x$ or $x = 1$

$\{1, 7\}$

14. $\{-2, 5\}$

15. $|2x + 11| = |3x - 4|$

$2x + 11 = 3x - 4$ or $2x + 11 = -(3x - 4)$
$\qquad\qquad\qquad\qquad 2x + 11 = -3x + 4$

$\quad 11 = x - 4$ or $5x + 11 = 4$

$\qquad\qquad\qquad\qquad 5x = -7$

$\quad 15 = x$ or $x = -1.4$

$\{-1.4, 15\}$

16. $\left\{-6, \dfrac{20}{9}\right\}$

17. $|2x - 3| - 7 = 6$

$\quad |2x - 3| = 13$

$2x - 3 = 13$ or $2x - 3 = -13$

$2x = 16$ or $2x = -10$

$x = 8$ or $x = -5$

$\{-5, 8\}$

18. $\{-7, 9.8\}$

19. $|-19| = |2x + 1|$

$\quad 19 = |2x + 1|$

$19 = 2x + 1$ or $-19 = 2x + 1$

$18 = 2x$ or $-20 = 2x$

$9 = x$ or $-10 = x$

$\{-10, 9\}$

20. $\left\{-5\dfrac{1}{2}, 7\dfrac{1}{6}\right\}$

21. $|x + 7| = |2x - 14|$

$x + 7 = 2x - 14$ or $x + 7 = -(2x - 14)$
$\qquad\qquad\qquad\qquad x + 7 = -2x + 14$

$\quad 7 = x - 14$ or $3x + 7 = 14$

$\qquad\qquad\qquad\qquad 3x = 7$

$\quad 21 = x$ or $x = \dfrac{7}{3}$

$\left\{\dfrac{7}{3}, 21\right\}$

22. $\left\{-7, -\dfrac{17}{7}\right\}$

23. $\left|\dfrac{x + 3}{4}\right| = 16$

$\dfrac{x + 3}{4} = 16$ or $\dfrac{x + 3}{4} = -16$

$x + 3 = 64$ or $x + 3 = -64$

$x = 61$ or $x = -67$

$\{-67, 61\}$

24. $\{-10, 11\}$

25. (a) $100\% - (12\% + 20\% + 17\% + 14\% + 2\% + 14\%)$

$= 100\% - 79\%$

$= 21\%$

(b) $12\% + 20\% = 32\%$

(c) 12% of $360 = 0.12(360) = 43.2°$

(d) 17% of $500 = 0.17(500) = 85$ people

Section 2.7

1. $|w| < 27$

$-27 < w < 27$

$(-27, 27)$

2. $[-6.5, 6.5]$

3. $|4 - 2x| \le 12$

$-12 \le 4 - 2x \le 12$

$-16 \le -2x \le 8$

$\dfrac{-16}{-2} \ge \dfrac{-2x}{-2} \ge \dfrac{8}{-2}$

$8 \ge x \ge -4$

$[-4, 8]$

4. $\left(-8, \dfrac{8}{3}\right)$

5. $|5x| \ge 45$

$5x \le -45$ or $5x \ge 45$

$x \le -9$ or $x \ge 9$

$(-\infty, -9] \cup [9, \infty)$

6. $\left(-\infty, -\dfrac{7}{2}\right) \cup \left(\dfrac{7}{2}, \infty\right)$

7. $|10x - 6| > 54$

$10x - 6 < -54$ or $10x - 6 > 54$

$10x < -48$ or $10x > 60$

$x < -4.8$ or $x > 6$

$(-\infty, -4.8) \cup (6, \infty)$

8. $(-\infty, -9] \cup [15, \infty)$

9. $|x + 2| > 23$

$x + 2 > 23$ or $x + 2 < -23$

$x > 21$ or $x < -25$

$(-\infty, -25) \cup (21, \infty)$

10. $(-\infty, -6] \cup \left[\dfrac{20}{3}, \infty\right)$

11. $|2x+11| \le -4$

Absolute value is always nonnegative and can never be less than –4. Thus, this inequality has no solution.

$\varnothing$

12. $\varnothing$

13. $|2x-3|-7 < 6$

$$|2x-3| < 13$$
$$-13 < 2x-3 < 13$$
$$-10 < 2x < 16$$
$$-5 < x < 8$$
$$(-5, 8)$$

14. $\{-9, 11.8\}$

15. $|2(x+3)| > 12$

$$2(x+3) > 12 \quad \text{or} \quad 2(x+3) < -12$$
$$x+3 > 6 \quad \text{or} \quad x+3 < -6$$
$$x > 3 \quad \text{or} \quad x < -9$$
$$(-\infty, -9) \cup (3, \infty)$$

16. $[1, 3]$

17. $\left|\dfrac{w-5}{3}\right| \ge 8$

$$\frac{w-5}{3} \ge 8 \quad \text{or} \quad \frac{w-5}{3} \le -8$$
$$w-5 \ge 24 \quad \text{or} \quad w-5 \le -24$$
$$w \ge 29 \quad \text{or} \quad w \le -19$$
$$(-\infty, -19] \cup [29, \infty)$$

18. $\left(-14\dfrac{2}{3}, 8\dfrac{2}{3}\right)$

19. $\left|\dfrac{63}{37} - X\right| \le 0.05$

$$-0.05 \le \frac{63}{37} - X \le 0.05$$
$$-1.7527\overline{027} \le -X \le -1.6527\overline{027}$$
$$1.7527\overline{027} \ge X \ge 1.6527\overline{027}$$
$$\left[1.6527\overline{027}, 1.7527\overline{027}\right]$$

20. $(0.078, 0.08)$

21.

(a) $P(A) = \dfrac{1}{5}$

(b) $P(P) = \dfrac{2}{5}$

(c) $P(\text{vowel}) = P(A \text{ or } E) = \dfrac{2}{5}$

(d) $P(\text{consonant}) = P(P \text{ or } L) = \dfrac{3}{5}$

Chapter 2 Practice Test

1. $\{-8\}$
2. $\{13\}$
3. $\{2\}$
4. $\{n \mid n \text{ is a real number}\}$
5. $\{26\}$
6. $\{-150\}$
7. $\{1, 4\}$
8. $\{\ \}$
9. $y = \dfrac{5x}{7} - 2$
10. $r = \dfrac{45q}{8}$
11. $v = \dfrac{S - gt^2}{gt}$
12. $w = \dfrac{V}{lh}$
13. $(-4, \infty)$
14. $[24, \infty)$
15. $[0, 3)$
16. $(-\infty, -2) \cup (5, \infty)$
17. $(-\infty, -6]$
18. $(-\infty, -1]$
19. $(3, 11]$
20. $(-\infty, \infty)$
21. 20.9
22. 42%
23. Approximately 40 inches by 40 inches
24. The company must sell more than 395 games to make a profit.
25. $6161.64

Chapter 3

Section 3.1

1. $(2, 5)$
$$2 + 5 = 8$$
$$7 = 8 \quad \text{False}$$
$(-1, 9)$
$$-1 + 9 = 8$$
$$8 = 8 \quad \text{True}$$
Thus, $(2, 5)$ is not a solution and $(-1, 9)$ is a solution.

2. Solution; not a solution

3. $(3, 26)$
$$-5(3) + 26 = 11$$
$$-15 + 26 = 11$$
$$11 = 11 \quad \text{True}$$
$(-2, 0)$

$$-5(-2) + 0 = 11$$
$$10 + 0 = 11$$
$$10 = 11 \quad \text{False}$$

Thus, (3, 26) is a solution and (−2, 0) is not a solution.

4. Not a solution; solution

5.

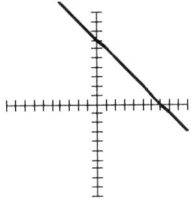

6.

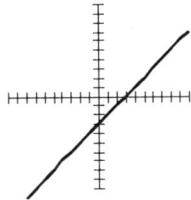

7.

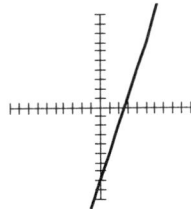

8.

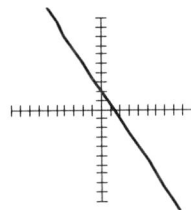

9.

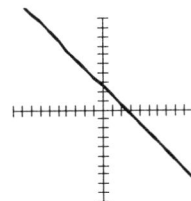

10.

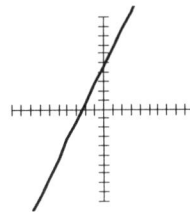

11.

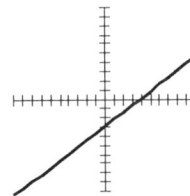

12.

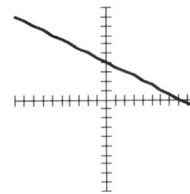

13.

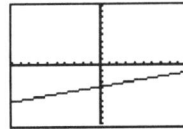

14.

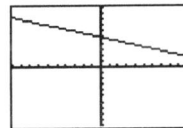

15.

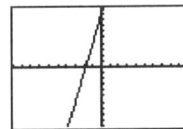

16.

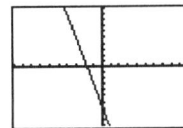

Section 3.2

1. Domain: {5, 17, 3, 6}
Range: {2, 0, −4}
Function

2. Domain: {1, 4, 3, 2}
Range: {3, −10, 5, 2}
Function

3. Domain: {1, 2, 3, 4}

Range: {2, 1, 3}
Not a function

4. Domain: {2, 1, 3, 4, 0}
Range: {2, 1, 4}
Function

5. Domain: {a, b, c, 2}
Range: {1, 4}
Function

6. Domain: {a, b, c, 5}
Range: {a, 3, 1, b}
Not a function

7. $f(4) = 5(4) - 4$
$= 16$

8. -14

9. $h(0) = 2(0)^2 + 1$
$= 1$

10. 3

11. $g(4) = 4 - (4)^2$
$= 4 - 16$
$= -12$

12. 0

13.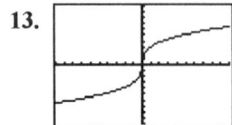
Function

14.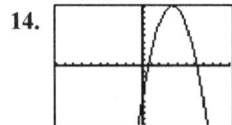
Function

15. Function

16. Not a function

Section 3.3

1.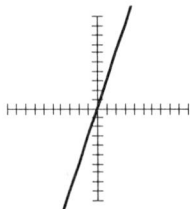

2.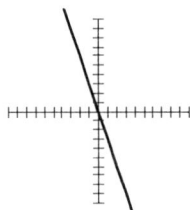

3.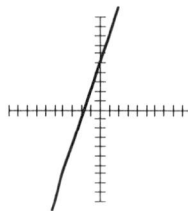

4.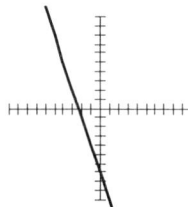

5.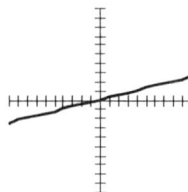

6.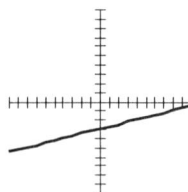

7.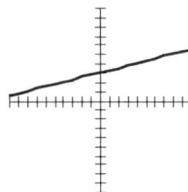

8.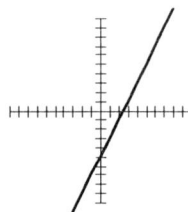

9. $0 + y = 7$ $\qquad$ $x + 0 = 7$
$y = 7$ $\qquad$ $x = 7$
$(0, 7), (7, 0)$

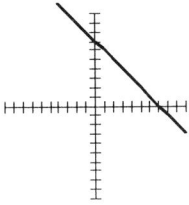

10. (0, 2), (1, 0)

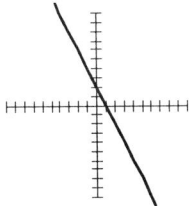

11. $0 + 2y = 5 \qquad x + 2(0) = 5$

$\qquad 2y = 5 \qquad\qquad x = 5$

$\qquad y = \dfrac{5}{2}$

$\qquad \left(0, \dfrac{5}{2}\right), (5, 0)$

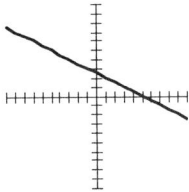

12. (0, 4), (4, 0)

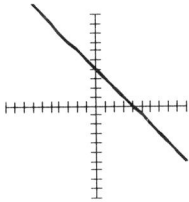

13. $0 - 5y = 10 \qquad x - 5(0) = 10$

$\qquad -5y = 10 \qquad\quad x - 0 = 10$

$\qquad\quad y = -2 \qquad\qquad x = 10$

$\quad (0, -2), (10, 0)$

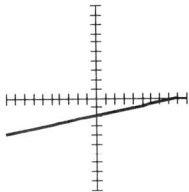

14. (0, 4), (–12, 0)

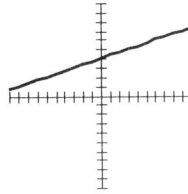

15. $6x + 9y = -8$

$\qquad 9y = -8 - 6x$

$\qquad y = -\dfrac{8}{9} - \dfrac{2}{3}x$

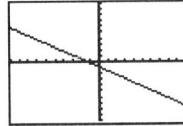

16. $y = \dfrac{7}{3}x - 4$

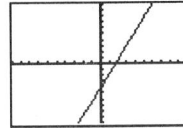

17. (d)
18. (a)
19. (b)
20. (c)
21. (a) $\quad y = 250(0) + 400$

$\qquad\quad y = 400$

(0, 400)

This means if you don't work at all, your checking account balance will remain at $400.

(b)

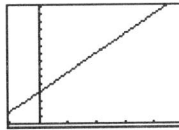

(c) $\quad y = 250(5) + 400 = 1650$

The balance after 5 weeks will be $1650.
22. (a) 3,600,700 teachers and librarians
(b) in 2007
(c) Answers will vary.

Section 3.4

1. $\quad m = \dfrac{5 - (-2)}{3 - 7} = \dfrac{7}{-4} = -\dfrac{7}{4}$

2. 3

3. $\quad m = \dfrac{1 - 3}{5 - (-12)} = \dfrac{-2}{17} = -\dfrac{2}{17}$

4. $-\dfrac{2}{3}$

5. $m = \dfrac{\dfrac{3}{4} - \dfrac{1}{4}}{-\dfrac{1}{2} - 4} = \dfrac{\dfrac{1}{2}}{-\dfrac{9}{2}} = -\dfrac{1}{9}$

6. 15

7. $x + y = 7$

$y = 7 - x$

$m = -1$, y-intercept $= 7$

8. $m = -2$, y-intercept $= 2$

9. $x + 2y = 5$

$2y = 5 - x$

$y = \dfrac{5}{2} - \dfrac{1}{2}x$

$m = -\dfrac{1}{2}$, y-intercept $= \dfrac{5}{2}$

10. $m = -1$, y-intercept $= 4$

11. $x - 5y = 10$

$-5y = 10 - x$

$y = -2 + \dfrac{1}{5}x$

$m = \dfrac{1}{5}$, y-intercept $= -2$

12. $m = \dfrac{1}{3}$, y-intercept $= 4$

13.

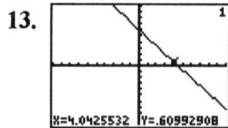

14.

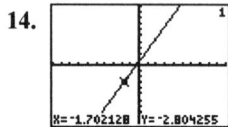

15. Neither

16. Perpendicular

17. $3x - y = 6$ $2x + 6y = 13$

$3x - 6 = y$ $6y = 13 - 2x$

$y = \dfrac{13}{6} - \dfrac{1}{3}x$

Perpendicular

18. Parallel

19. (1992, 73,449) and (1994, 61,775)

$m = \dfrac{73,449 - 61,775}{1992 - 1994} = \dfrac{11,674}{-2} = -5837$

20. Approximately -2.38

Section 3.5

1. $m = 2$, $b = 0$

$y = 2x + 0$

$y = 2x$

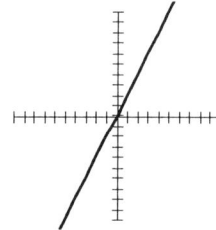

2. $y = -3x + 2$

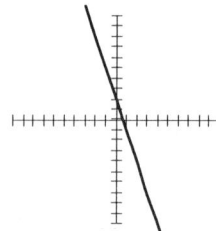

3. $m = \dfrac{1}{2}$, $b = -3$

$y = \dfrac{1}{2}x - 3$

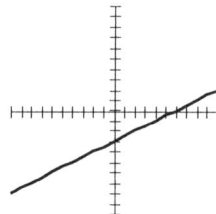

4. $y = -\dfrac{4}{3}x - 1$

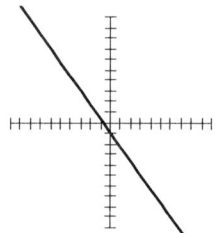

5. $m = \dfrac{5 - (-2)}{3 - 7} = -\dfrac{7}{4}$

$$y - 5 = -\frac{7}{4}(x - 3)$$

$$4y - 20 = -7x + 21$$

$$7x + 4y = 41$$

6. $4x + 36y = 25$

7. $$m = \frac{1 - 3}{5 - (-12)} = \frac{-2}{17} = -\frac{2}{17}$$

$$y - 1 = -\frac{2}{17}(x - 5)$$

$$17y - 17 = -2x + 10$$

$$2x + 17y = 27$$

8. $$f(x) = -\frac{2}{3}x + \frac{1}{3}$$

9. $$m = \frac{18 - 0}{6 - 0} = 3$$

$$f(x) = 3x$$

10. $f(x) = 15x - 7$

11. $x + y = 7 \implies m = -1$

$$y - 1 = -1(x - 1)$$

$$y - 1 = -x + 1$$

$$f(x) = -x + 2$$

12. $$f(x) = \frac{1}{5}x + \frac{33}{5}$$

13. $x + 2y = 5 \implies m = -\frac{1}{2}$

$$m = 2$$

$$y + 4 = 2(x + 1)$$

$$y + 4 = 2x + 2$$

$$f(x) = 2x - 2$$

14. $f(x) = -3x + 1$

15. $(2, \ 26{,}057)$ and $(4, \ 24{,}084)$

$$m = \frac{26{,}057 - 24{,}084}{2 - 4} = -986.5$$

$$y - 26{,}057 = -986.5(x - 2)$$

$$y - 26{,}057 = -986.5x + 1973$$

$$y = -986.5x + 28{,}030$$

16. $y = 1.54x + 9$

Section 3.6

1.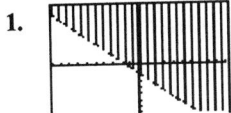

2.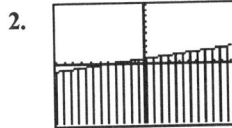

3.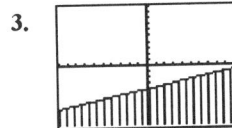

4.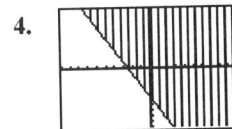

5.

6.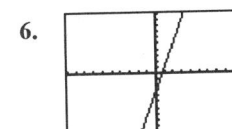

7.

8.

9. (d)
10. (a)
11. (c)
12. (b)
13. $x \geq 12$ and $y \leq 25$

The points in the intersection of these two inequalities are the sets of all times such that the time spent baking is at least 12 hours and the time spent fishing is no more than 25 hours.

Chapter 3 Practice Test

1. II, I, IV, III
2. $(-24, -4)$
3.

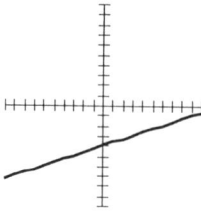

4.

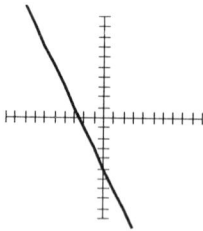

5.

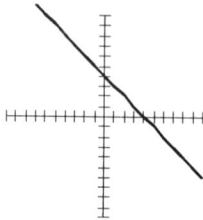

6.

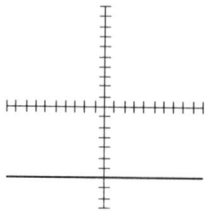

7. $\dfrac{9}{2}$
8. $m = 4, b = -3$
9.

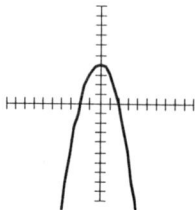

10.

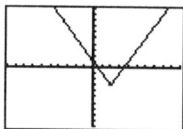

11. $y = 3$
12. $x = 5$
13. $x = -5$
14. $y = -1$
15. $y = -2x + 6$
16. $y = 4x - 48$
17. $y = -x + 6$
18. $y = 2x + 9$
19.

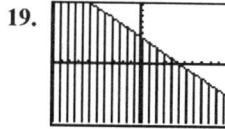

20.

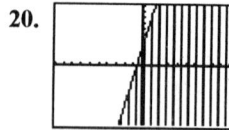

21. Domain: $\{0, 3, -2, 1\}$
    Range: $\{7, -3, 4, 11\}$
    Not a function
22. Domain: $\{2, 1, 3, -1, -3\}$
    Range: $\{-6, 12, 4\}$
    Function
23. (a) $f(5) = 10{,}184$
    (b) $11{,}241$
    (c) 1999

Chapter 4

Section 4.1

1.

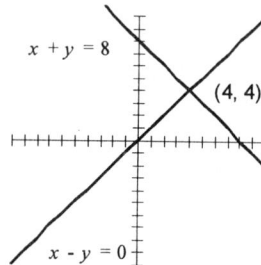

The solution set is $\{(4, 4)\}$.

2. $\{(3, 6)\}$

3.

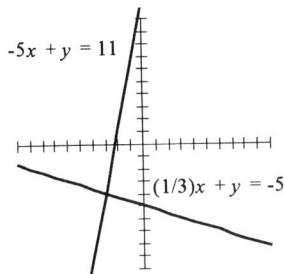

The solution set is $\{(-3, -4)\}$

4. $\left\{\left(\dfrac{8}{3}, \dfrac{40}{9}\right)\right\}$

5. $x - 2y = 8$

$x = 2y + 8$

$3x + 4y = 16$

$3(2y + 8) + 4y = 16$

$10y + 24 = 16$

$y = -\dfrac{8}{10} = -\dfrac{4}{5}$

$x = 2\left(-\dfrac{4}{5}\right) + 8 = \dfrac{32}{5}$

The solution set is $\left\{\left(\dfrac{32}{5}, -\dfrac{4}{5}\right)\right\}$.

6. $\{(0.519, -2.192)\}$

7. $\begin{cases} 3x - 2y = 9 \\ 5x + 2y = 7 \end{cases}$

$8x = 16$

$x = 2$

$3x - 2y = 9$

$3(2) - 2y = 9$

$y = -\dfrac{3}{2}$

The solution set is $\left\{\left(2, -1.5\right)\right\}$.

8. $\{(1.531, -4.107)\}$

9. $\begin{cases} -2x - 5y = 11 \\ 3x - 3y = 17 \end{cases}$

$\begin{cases} -6x - 15y = 33 \\ 6x - 6y = 34 \end{cases}$

$-21y = 67$

$y = -\dfrac{67}{21} \approx -3.190$

$3x - 3y = 17$

$3x - 3\left(-\dfrac{67}{21}\right) = 17$

$x = \dfrac{52}{21} \approx 2.476$

The solution set is $\{(2.476, -3.190)\}$

10. $\{(3.13, -3.94)\}$

11. $\{(-1.38, -4.08)\}$

12. $\{(14.14, -2.41)\}$

13. $\{(-4.41, -5.17)\}$

14. $\{(2.429, -26.143)\}$

15. $\begin{cases} x + 13y = 26 \\ -8x + y = -15 \end{cases}$

$\begin{cases} 8x + 104y = 208 \\ -8x + y = -15 \end{cases}$

$105y = 193$

$y = \dfrac{193}{105} \approx 1.838$

$-8x + y = -15$

$-8x + \dfrac{193}{105} = -15$

$x = \dfrac{1768}{840} \approx 2.105$

The solution set is $\{(2.105, 1.838)\}$

16. $\{(2, 2)\}$

17. $\begin{cases} x + 2y = 20 \\ 2x - \dfrac{y}{3} = 10 \end{cases}$

$\begin{cases} x + 2y = 20 \\ 12x - 2y = 60 \end{cases}$

$13x = 80$

$x = \dfrac{80}{13} \approx 6.154$

$x + 2y = 20$

$\dfrac{80}{13} + 2y = 20$

$y = \dfrac{90}{13} \approx 6.923$

The solution set is $\{(6.154, 6.923)\}$

18. $\{(-11.622, 0.946)\}$

19. $\begin{cases} \dfrac{x}{12} - \dfrac{5y}{3} = \dfrac{7}{4} \\ 6x + 120y = -126 \end{cases}$

$\begin{cases} x - 20y = 21 \\ x + 20y = -21 \end{cases}$

$\overline{ x = 0}$

$x - 20y = 21$

$0 - 20y = 21$

$y = -\dfrac{21}{20} = -1.05$

The solution set is $\{(0, -1.05)\}$

20. 1482 quarts

21. $y = 16x$

$y = 4.25x + 7500$

$16x = 4.25x + 7500$

$11.75x = 7500$

$x = \dfrac{7500}{11.75}$

The salon must give 639 haircuts to break even.

22. (a) y_1 is the cost equation; y_2 is the revenue equation

(b) Approximately (300, 750)

Section 4.2

1. $\begin{cases} x + z = -3 \\ 5y = 15 \\ x + y - z = 4 \end{cases}$

$x + 3 - z = 4$

$5y = 15 \qquad x - z = 1$

$y = 3 \qquad \dfrac{x + z = -3}{2x = -2}$

$ x = -1$

$x + z = -3$

$-1 + z = -3$

$z = -2$

The solution set is $\{(-1, 3, -2)\}$

2. $\{(13, 10.364, -3.545)\}$

3. $\begin{cases} x + y + z = 0 \\ -2y + \dfrac{1}{2}z = 4 \\ -x + y = 12 \end{cases}$

$x + y + z = 0$

$\dfrac{-x + y = 12}{2y + z = 12}$

$-2y + \dfrac{1}{2}z = 4$

$\overline{\dfrac{3}{2}z = 16}$

$z = \dfrac{32}{3}$

$-2y + \dfrac{1}{2}z = 4$

$-2y + \left(\dfrac{1}{2}\right)\left(\dfrac{32}{3}\right) = 4$

$y = \dfrac{2}{3}$

$x + y + z = 0$

$x = -\dfrac{2}{3} - \dfrac{32}{3} = -\dfrac{34}{3}$

The solution set is $\left\{\left(-\dfrac{34}{3}, \dfrac{2}{3}, \dfrac{32}{3}\right)\right\}$.

4. $\{(3.435, 2.304, 0.783)\}$

5. $\begin{cases} x - y + 2z = 0 \\ x + 2y + z = 0 \\ 2x - y - z = 0 \end{cases}$

$2x - 2y + 4z = 0 \qquad\qquad x + 2y + z = 0$

$\dfrac{x + 2y + z = 0}{3x + 5z = 0} \qquad\qquad \dfrac{4x - 2y - 2z = 0}{5x - z = 0}$

$3x + 5z = 0$

$\dfrac{25x - 5z = 0}{28x = 0}$

$x = 0$

$3(0) + 5z = 0$

$5z = 0$

$z = 0$

$x - y + 2z = 0$

$0 - y + 2(0) = 0$

$-y = 0$

$y = 0$

The solution set is $\{(0, 0, 0)\}$.

6. $\{(-1.5, -1.333, 3.833)\}$
7. One example is:
$$\begin{cases} x + y + z = 0 \\ \quad\quad y \quad\quad = 0 \\ \quad\quad\quad\quad z = 0 \end{cases}$$
8. One example is:
$$\begin{cases} 2x + \quad 10z = \quad 52 \\ \quad\quad y \quad\quad\quad = -10 \\ \quad\quad y + \quad z = \quad -5 \end{cases}$$

Section 4.3

1. 37.5 minutes is 0.625 hours
$$0.625x + 0.625y = 50$$
$$y = 3x$$
$$0.625x + 0.625(3x) = 50$$
$$2.5x = 50$$
$$x = 20$$
The bicycle and car meet $0.625(20) = 12.5$ miles from Morris.
2. 183 miles from Lubbock
3. $x = y - 6$
$$\frac{1}{2}x = 4y + 25$$
$$\frac{1}{2}(y - 6) = 4y + 25$$
$$y - 6 = 8y + 50$$
$$-56 = 7y$$
$$y = -8$$
The two numbers are –8 and –14.
4. 19, 76
5. $x + y = 185$
$$\underline{x - y = 171}$$
$$2x = 356$$
$$x = 178$$
$$x + y = 185$$
$$178 + y = 185$$
$$y = 7$$
The speed of the car is 178 mph and the speed of the wind is 7 mph.
6. Speed of team in still water: 18.1 km/h
Speed of current: 2.1 km/h
7.
$$12x + 6y = 6.90$$
$$6x + 12y = 7.50$$
$$12x + 6y = 6.90$$
$$\underline{-12x - 24y = -15}$$
$$-18y = -8.10$$
$$y = 0.45$$

$$12x + 6y = 6.90$$
$$12x + 6(0.45) = 6.90$$
$$x = 0.35$$
An apple costs $0.35 and an orange costs $0.45.
8. Impatiens cost $1.69 per flat, marigolds cost $0.99 per flat
9. $x + 3(y - 5) + y = 210$
$$x = \frac{1}{2}y$$
$$\frac{1}{2}y + 3(y - 5) + y = 210$$
$$\frac{1}{2}y + 3y - 15 + y = 210$$
$$\frac{9}{2}y = 225$$
$$y = 50$$
$$x = \frac{1}{2}y = \frac{1}{2}(50) = 25$$
$$y - 5 = 50 - 5 = 45$$
The sides are 50, 45, 45, 45, and 25 centimeters.
10. The sides are 18, 18, and 32 inches.
11. $y = 50 + 0.3x$
$$y = 40 + 0.45x$$
$$50 + 0.3x = 40 + 0.45x$$
$$10 = 0.15x$$
$$66.67 \approx x$$
The total cost of renting a truck is the same at both rental companies for driving approximately 66.67 miles.
12. Cost is the same for 15 cubic feet of helium
13. $R(x) = 38x$
$$C(x) = 13x + 300$$
$$38x = 13x + 300$$
$$25x = 300$$
$$x = 12$$
Her break even point is 12 plant stands.
14. The florist must sell 40.2 dozen to break even.
15. $x + y + z = 100$
$$z = 3x$$
$$y = 2z - 12$$
$$x + y + z = 100$$
$$x + 2z - 12 + 3x = 100$$
$$x + 2(3x) - 12 + 3x = 100$$
$$10x = 112$$
$$x = 11.2$$

$z = 3x = 33.6$

$y = 2z - 12 = 55.2$

The numbers are 11.2, 33.6, and 55.2.

16. $-\dfrac{50}{3}, 0, \dfrac{50}{3}$

17.
$$a + b + c = 45$$
$$49a + 7b + c = 171$$
$$169a + 13b + c = 153$$

$$49a + 7b + c = 171$$
$$\underline{-7a - 7b - 7c = -315}$$
$$42a - 6c = -144$$

$$169a + 13b + c = 153$$
$$\underline{-13a - 13b - 13c = -585}$$
$$156a - 12c = -432$$

$$-2(42a - 6c = -144)$$
$$\underline{156a - 12c = -432}$$
$$72a = -144$$
$$a = -2$$
$$42(-2) - 6c = -144$$
$$-6c = -60$$
$$c = 10$$
$$-2 + b + 10 = 45$$
$$b = 37$$

The model is $y = -2x^2 + 37x + 10$.

18. $y = 0.18x^2 - 1.5x + 6.0$

19. $x \approx 89$ units

20. $x \approx 139$ units

Section 4.4

1. $\begin{bmatrix} 1 & 1 & \vdots & -10 \\ 1 & -6 & \vdots & 18 \end{bmatrix}$

$\begin{bmatrix} 1 & 1 & \vdots & -10 \\ -1(1)+1 & -1(1)-6 & \vdots & -1(-10)+18 \end{bmatrix}$

$\begin{bmatrix} 1 & 1 & \vdots & -10 \\ 0 & -7 & \vdots & 28 \end{bmatrix}$

$\begin{bmatrix} 1 & 1 & \vdots & -10 \\ 0 \cdot \dfrac{-1}{7} & -7 \cdot \dfrac{-1}{7} & \vdots & 28 \cdot \dfrac{-1}{7} \end{bmatrix}$

$\begin{bmatrix} 1 & 1 & \vdots & -10 \\ 0 & 1 & \vdots & -4 \end{bmatrix}$

$x + y = -10$

$y = -4$

$x + (-4) = -10$

$x = -6$

The solution set is $\{(-6, -4)\}$.

2. $\{(15.75, 2.25)\}$

3. $\begin{bmatrix} 1 & -5 & \vdots & 15 \\ 2 & 5 & \vdots & 12 \end{bmatrix}$

$\begin{bmatrix} 1 & -5 & \vdots & 15 \\ -2(1)+2 & -2(-5)+5 & \vdots & -2(15)+12 \end{bmatrix}$

$\begin{bmatrix} 1 & -5 & \vdots & 15 \\ 0 & 15 & \vdots & -18 \end{bmatrix}$

$\begin{bmatrix} 1 & -5 & \vdots & 15 \\ \dfrac{0}{15} & \dfrac{15}{15} & \vdots & \dfrac{-18}{15} \end{bmatrix}$

$\begin{bmatrix} 1 & -5 & \vdots & 15 \\ 0 & 1 & \vdots & -1.2 \end{bmatrix}$

$x - 5y = 15$

$y = -1.2$

$x - 5(-1.2) = 15$

$x = 9$

The solution set is $\{(9, -1.2)\}$.

4. $\{(-2, -2)\}$

5. $\begin{bmatrix} 3 & -4 & \vdots & 20 \\ -9 & 12 & \vdots & 11 \end{bmatrix}$

$\begin{bmatrix} 3 & -4 & \vdots & 20 \\ 3(3)-9 & 3(-4)+12 & \vdots & 3(20)+11 \end{bmatrix}$

$\begin{bmatrix} 3 & -4 & \vdots & 20 \\ 0 & 0 & \vdots & 71 \end{bmatrix}$

The system is inconsistent and has no solution.

6. No solution

7. $\{(0.2, 8.4)\}$

8. $\{(-2.75, 2.5)\}$

9. $\left\{ \left(\dfrac{10}{3}, -\dfrac{35}{6} \right) \right\}$

10. $\{(12, -14, 2)\}$

11. $\{(-1, 8, -6)\}$

12. $\{(20.5, -29.5, 41)\}$

13. $\{(5.714, -4.143, 4.857)\}$

14. $\{(-49, 74, -8)\}$

15. $\{(-0.364, 6.273)\}$

16. $\{(2.044, -1.215)\}$

6. $\{(-1.5, -1.333, 3.833)\}$
7. One example is:
$$\begin{cases} x+y+z=0 \\ \quad\;\; y \quad\;\; = 0 \\ \quad\quad\;\; z = 0 \end{cases}$$

8. One example is:
$$\begin{cases} 2x + \quad 10z = \quad 52 \\ \quad\; y \quad\quad\; = -10 \\ \quad\; y + \; z = \; -5 \end{cases}$$

Section 4.3

1. 37.5 minutes is 0.625 hours
$$0.625x + 0.625y = 50$$
$$y = 3x$$
$$0.625x + 0.625(3x) = 50$$
$$2.5x = 50$$
$$x = 20$$
The bicycle and car meet $0.625(20) = 12.5$ miles from Morris.

2. 183 miles from Lubbock
3. $x = y - 6$
$$\frac{1}{2}x = 4y + 25$$
$$\frac{1}{2}(y - 6) = 4y + 25$$
$$y - 6 = 8y + 50$$
$$-56 = 7y$$
$$y = -8$$
The two numbers are –8 and –14.

4. 19, 76
5. $x + y = 185$
$$\underline{x - y = 171}$$
$$2x = 356$$
$$x = 178$$
$$x + y = 185$$
$$178 + y = 185$$
$$y = 7$$
The speed of the car is 178 mph and the speed of the wind is 7 mph.

6. Speed of team in still water: 18.1 km/h
Speed of current: 2.1 km/h

7.
$$12x + 6y = 6.90$$
$$12x + 6y = 6.90 \qquad \underline{-12x - 24y = -15}$$
$$6x + 12y = 7.50 \qquad -18y = -8.10$$
$$y = 0.45$$

$$12x + 6y = 6.90$$
$$12x + 6(0.45) = 6.90$$
$$x = 0.35$$
An apple costs $0.35 and an orange costs $0.45.

8. Impatiens cost $1.69 per flat, marigolds cost $0.99 per flat

9. $x + 3(y - 5) + y = 210$
$$x = \frac{1}{2}y$$
$$\frac{1}{2}y + 3(y - 5) + y = 210$$
$$\frac{1}{2}y + 3y - 15 + y = 210$$
$$\frac{9}{2}y = 225$$
$$y = 50$$
$$x = \frac{1}{2}y = \frac{1}{2}(50) = 25$$
$$y - 5 = 50 - 5 = 45$$
The sides are 50, 45, 45, 45, and 25 centimeters.

10. The sides are 18, 18, and 32 inches.

11. $y = 50 + 0.3x$
$$y = 40 + 0.45x$$
$$50 + 0.3x = 40 + 0.45x$$
$$10 = 0.15x$$
$$66.67 \approx x$$
The total cost of renting a truck is the same at both rental companies for driving approximately 66.67 miles.

12. Cost is the same for 15 cubic feet of helium

13. $R(x) = 38x$
$$C(x) = 13x + 300$$
$$38x = 13x + 300$$
$$25x = 300$$
$$x = 12$$
Her break even point is 12 plant stands.

14. The florist must sell 40.2 dozen to break even.

15. $x + y + z = 100$
$$z = 3x$$
$$y = 2z - 12$$
$$x + y + z = 100$$
$$x + 2z - 12 + 3x = 100$$
$$x + 2(3x) - 12 + 3x = 100$$
$$10x = 112$$
$$x = 11.2$$

$z = 3x = 33.6$

$y = 2z - 12 = 55.2$

The numbers are 11.2, 33.6, and 55.2.

16. $-\dfrac{50}{3}$, 0, $\dfrac{50}{3}$

17.

$a + b + c = 45$

$49a + 7b + c = 171$

$169a + 13b + c = 153$

$\begin{array}{r} 49a + 7b + c = 171 \\ -7a - 7b - 7c = -315 \\ \hline 42a - 6c = -144 \end{array}$

$\begin{array}{r} 169a + 13b + c = 153 \\ -13a - 13b - 13c = -585 \\ \hline 156a - 12c = -432 \end{array}$

$\begin{array}{r} -2(42a - 6c = -144) \\ 156a - 12c = -432 \\ \hline 72a = -144 \\ a = -2 \end{array}$

$42(-2) - 6c = -144$

$-6c = -60$

$c = 10$

$-2 + b + 10 = 45$

$b = 37$

The model is $y = -2x^2 + 37x + 10$.

18. $y = 0.18x^2 - 1.5x + 6.0$

19. $x \approx 89$ units

20. $x \approx 139$ units

Section 4.4

1. $\begin{bmatrix} 1 & 1 & \vdots & -10 \\ 1 & -6 & \vdots & 18 \end{bmatrix}$

$\begin{bmatrix} 1 & 1 & \vdots & -10 \\ -1(1)+1 & -1(1)-6 & \vdots & -1(-10)+18 \end{bmatrix}$

$\begin{bmatrix} 1 & 1 & \vdots & -10 \\ 0 & -7 & \vdots & 28 \end{bmatrix}$

$\begin{bmatrix} 1 & 1 & \vdots & -10 \\ 0 \cdot \dfrac{-1}{7} & -7 \cdot \dfrac{-1}{7} & \vdots & 28 \cdot \dfrac{-1}{7} \end{bmatrix}$

$\begin{bmatrix} 1 & 1 & \vdots & -10 \\ 0 & 1 & \vdots & -4 \end{bmatrix}$

$x + y = -10$

$y = -4$

$x + (-4) = -10$

$x = -6$

The solution set is $\{(-6, -4)\}$.

2. $\{(15.75, 2.25)\}$

3. $\begin{bmatrix} 1 & -5 & \vdots & 15 \\ 2 & 5 & \vdots & 12 \end{bmatrix}$

$\begin{bmatrix} 1 & -5 & \vdots & 15 \\ -2(1)+2 & -2(-5)+5 & \vdots & -2(15)+12 \end{bmatrix}$

$\begin{bmatrix} 1 & -5 & \vdots & 15 \\ 0 & 15 & \vdots & -18 \end{bmatrix}$

$\begin{bmatrix} 1 & -5 & \vdots & 15 \\ \dfrac{0}{15} & \dfrac{15}{15} & \vdots & \dfrac{-18}{15} \end{bmatrix}$

$\begin{bmatrix} 1 & -5 & \vdots & 15 \\ 0 & 1 & \vdots & -1.2 \end{bmatrix}$

$x - 5y = 15$

$y = -1.2$

$x - 5(-1.2) = 15$

$x = 9$

The solution set is $\{(9, -1.2)\}$.

4. $\{(-2, -2)\}$

5. $\begin{bmatrix} 3 & -4 & \vdots & 20 \\ -9 & 12 & \vdots & 11 \end{bmatrix}$

$\begin{bmatrix} 3 & -4 & \vdots & 20 \\ 3(3)-9 & 3(-4)+12 & \vdots & 3(20)+11 \end{bmatrix}$

$\begin{bmatrix} 3 & -4 & \vdots & 20 \\ 0 & 0 & \vdots & 71 \end{bmatrix}$

The system is inconsistent and has no solution.

6. No solution

7. $\{(0.2, 8.4)\}$

8. $\{(-2.75, 2.5)\}$

9. $\left\{ \left(\dfrac{10}{3}, -\dfrac{35}{6} \right) \right\}$

10. $\{(12, -14, 2)\}$

11. $\{(-1, 8, -6)\}$

12. $\{(20.5, -29.5, 41)\}$

13. $\{(5.714, -4.143, 4.857)\}$

14. $\{(-49, 74, -8)\}$

15. $\{(-0.364, 6.273)\}$

16. $\{(2.044, -1.215)\}$

Section 4.5

1. $\begin{vmatrix} 4 & 7 \\ 3 & 7 \end{vmatrix} = (4)(7) - (3)(7) = 7$

2. 44

3. $\begin{vmatrix} 2 & -3 \\ 6 & 11 \end{vmatrix} = (2)(11) - (6)(-3) = 40$

4. −42

5. $\begin{vmatrix} 3 & -7 & 0 \\ 2 & 6 & -1 \\ -5 & 1 & 0 \end{vmatrix} = 0 - (-1) \cdot \begin{vmatrix} 3 & -7 \\ -5 & 1 \end{vmatrix} + 0 = -32$

6. 4

7. $\begin{vmatrix} 10 & 1 & 2 \\ 9 & 0 & 3 \\ 4 & -1 & -5 \end{vmatrix} = -9 \cdot \begin{vmatrix} 1 & 2 \\ -1 & -5 \end{vmatrix} + 0 - 3 \cdot \begin{vmatrix} 10 & 1 \\ 4 & -1 \end{vmatrix}$

$= (-9)[(1)(-5) - (-1)(2)] + 0 - (3)[(10)(-1) - (4)(1)]$

$= 69$

8. 112

9. $D = \begin{vmatrix} -1 & 5 \\ 14 & -9 \end{vmatrix}$

$D_x = \begin{vmatrix} 36 & 5 \\ 18 & -9 \end{vmatrix}$

$D_y = \begin{vmatrix} -1 & 36 \\ 14 & 18 \end{vmatrix}$

10. $D = \begin{vmatrix} 4 & 17 \\ 15 & -1 \end{vmatrix}$

$D_x = \begin{vmatrix} 100 & 17 \\ 76 & -1 \end{vmatrix}$

$D_y = \begin{vmatrix} 4 & 100 \\ 15 & 76 \end{vmatrix}$

11. $D = \begin{vmatrix} -81 & -37 \\ 18 & -11 \end{vmatrix}$

$D_x = \begin{vmatrix} 6 & -37 \\ 36 & -11 \end{vmatrix}$

$D_y = \begin{vmatrix} -81 & 6 \\ 18 & 36 \end{vmatrix}$

12. $D = \begin{vmatrix} -4 & 3 & -9 \\ 2 & -1 & 7 \\ 4 & 10 & -3 \end{vmatrix}$

$D_x = \begin{vmatrix} 21 & 3 & -9 \\ 58 & -1 & 7 \\ 25 & 10 & -3 \end{vmatrix}$

$D_y = \begin{vmatrix} -4 & 21 & -9 \\ 2 & 58 & 7 \\ 4 & 25 & -3 \end{vmatrix}$

$D_z = \begin{vmatrix} -4 & 3 & 21 \\ 2 & -1 & 58 \\ 4 & 10 & 25 \end{vmatrix}$

13. $D = \begin{vmatrix} 50 & -61 & 2 \\ -7 & 89 & 93 \\ 35 & -32 & -85 \end{vmatrix}$

$D_x = \begin{vmatrix} 12 & -61 & 2 \\ -41 & 89 & 93 \\ 21 & -32 & -85 \end{vmatrix}$

$D_y = \begin{vmatrix} 50 & 12 & 2 \\ -7 & -41 & 93 \\ 35 & 21 & -85 \end{vmatrix}$

$D_z = \begin{vmatrix} 50 & -61 & 12 \\ -7 & 89 & -41 \\ 35 & -32 & 21 \end{vmatrix}$

14. $D = \begin{vmatrix} 80 & 88 & 30 \\ -3 & 93 & -29 \\ 73 & -95 & 0 \end{vmatrix}$

$D_x = \begin{vmatrix} 67 & 88 & 30 \\ 38 & 93 & -29 \\ 40 & -95 & 0 \end{vmatrix}$

$D_y = \begin{vmatrix} 80 & 67 & 30 \\ -3 & 38 & -29 \\ 73 & 40 & 0 \end{vmatrix}$

$D_z = \begin{vmatrix} 80 & 88 & 67 \\ -3 & 93 & 38 \\ 73 & -95 & 40 \end{vmatrix}$

15. $D = \begin{vmatrix} 1 & 1 \\ 1 & -6 \end{vmatrix} = -7$

$D_x = \begin{vmatrix} -10 & 1 \\ 18 & -6 \end{vmatrix} = 42$

$D_y = \begin{vmatrix} 1 & -10 \\ 1 & 18 \end{vmatrix} = 28$

$x = \dfrac{42}{-7} = -6$

$y = \dfrac{28}{-7} = -4$

The solution set is {(−6, −4)}

16. {(15.75, 2.25)}

17. $D = \begin{vmatrix} 1 & -5 \\ 2 & 5 \end{vmatrix} = 15$

$D_x = \begin{vmatrix} 15 & -5 \\ 12 & 5 \end{vmatrix} = 135$

$D_y = \begin{vmatrix} 1 & 15 \\ 2 & 12 \end{vmatrix} = -18$

$x = \dfrac{135}{15} = 9$

$y = \dfrac{-18}{15} = -\dfrac{6}{5} = -1.2$

The solution set is {(9, −1.2)}

18. {(−2, −2)}

19.

$$D = \begin{vmatrix} 1 & 1 & 1 \\ 2 & 1 & 0 \\ 0 & 0 & 1 \end{vmatrix} = 0 - 0 + 1 \cdot \begin{vmatrix} 1 & 1 \\ 2 & 1 \end{vmatrix} = -1$$

$$D_x = \begin{vmatrix} 0 & 1 & 1 \\ 10 & 1 & 0 \\ 2 & 0 & 1 \end{vmatrix} = 0 - 1 \cdot \begin{vmatrix} 10 & 0 \\ 2 & 1 \end{vmatrix} + 1 \cdot \begin{vmatrix} 10 & 1 \\ 2 & 0 \end{vmatrix} = -12$$

$$D_y = \begin{vmatrix} 1 & 0 & 1 \\ 2 & 10 & 0 \\ 0 & 2 & 1 \end{vmatrix} = 1 \cdot \begin{vmatrix} 10 & 0 \\ 2 & 1 \end{vmatrix} - 0 + 1 \cdot \begin{vmatrix} 2 & 10 \\ 0 & 2 \end{vmatrix} = 14$$

$$D_z = \begin{vmatrix} 1 & 1 & 0 \\ 2 & 1 & 10 \\ 0 & 0 & 2 \end{vmatrix} = 0 - 0 + 2 \cdot \begin{vmatrix} 1 & 1 \\ 2 & 1 \end{vmatrix} = -2$$

$$x = \frac{-12}{-1} = 12$$

$$y = \frac{14}{-1} = -14$$

$$z = \frac{-2}{-1} = 2$$

The solution set is $\{(12, -14, 2)\}$

20. $\{(-1, 8, -6)\}$

21.

$$D = \begin{vmatrix} 1 & -1 & -1 \\ 1 & 1 & 1 \\ 2 & 0 & -1 \end{vmatrix} = 2 \cdot \begin{vmatrix} -1 & -1 \\ 1 & 1 \end{vmatrix} - 0 + (-1) \cdot \begin{vmatrix} 1 & -1 \\ 1 & 1 \end{vmatrix} = -2$$

$$D_x = \begin{vmatrix} 9 & -1 & -1 \\ 32 & 1 & 1 \\ 0 & 0 & -1 \end{vmatrix} = 0 - 0 + (-1) \cdot \begin{vmatrix} 9 & -1 \\ 32 & 1 \end{vmatrix} = -41$$

$$D_y = \begin{vmatrix} 1 & 9 & -1 \\ 1 & 32 & 1 \\ 2 & 0 & -1 \end{vmatrix} = 2 \cdot \begin{vmatrix} 9 & -1 \\ 32 & 1 \end{vmatrix} - 0 + (-1) \cdot \begin{vmatrix} 1 & 9 \\ 1 & 32 \end{vmatrix} = 59$$

$$D_z = \begin{vmatrix} 1 & -1 & 9 \\ 1 & 1 & 32 \\ 2 & 0 & 0 \end{vmatrix} = 2 \cdot \begin{vmatrix} -1 & 9 \\ 1 & 32 \end{vmatrix} - 0 + 0 = -82$$

$$x = \frac{-41}{-2} = 20.5$$

$$y = \frac{59}{-2} = -29.5$$

$$z = \frac{-82}{-2} = 41$$

The solution set is $\{(20.5, -29.5, 41)\}$.

22. $\{(5.714, -4.143, 4.857)\}$
23. $\{(1.48, 4.00)\}$
24. $\{(0.87, -0.28)\}$

Chapter 4 Test

1. -16
2. -582
3. $\{(-1, -2)\}$
4. $\{\ \}$
5. $\{(3, 0)\}$
6. $\{(x, y) \mid 7x - 7y = 15\}$

7. $\{(4, 0, -2)\}$
8. $\{(-6, 5, 8)\}$
9. $\left\{\left(-\frac{4}{5}, 8\right)\right\}$
10. $\{(5, -6)\}$
11. $\{(2, -4)\}$
12. $\{(-4, 5, -5)\}$
13. $\{(7, 6, 0)\}$
14. $\{(x, y) \mid x + 5y = 7\}$
15. $\{(-7, -8)\}$
16. $\{(2, -7, 6)\}$
17. $\{(-1, 5, 7)\}$
18. 500 jerseys
19. 285 rooms rented during the week and 160 rooms rented on the weekend
20. Approximately 16.4 pounds

Chapter 5

Section 5.1

1. $5^2 \cdot 5 = 5^2 \cdot 5^1 = 5^{2+1} = 5^3 = 125$
2. -32
3. $h^2 \cdot h^6 = h^{2+6} = h^8$
4. y^{12}
5. $x^3 \cdot x = x^3 \cdot x^1 = x^{3+1} = x^4$
6. p^6
7. $2^0 = 1$
8. 1
9. $q^0 = 1$
10. -4
11. $(-3)^{-3} = \frac{1}{(-3)^3} = -\frac{1}{27}$
12. $\frac{1}{625}$
13. $4a^{-2}a^{-3} = 4a^{-2-3} = 4a^{-5} = \frac{4}{a^5}$
14. 30
15. $a^5 b^{-5}(a^{-7}b^{-3}) = a^{5-7}b^{-5-3} = a^{-2}b^{-8} = \frac{1}{a^2 b^8}$
16. $\frac{5x^6}{y^2}$
17. $\frac{(4w^{-1})w^5}{w^8} = 4w^{-1+5-8} = 4w^{-4} = \frac{4}{w^4}$
18. $\frac{8q}{r^3}$
19. 5.744×10^9

20. 3.32×10^4

21. 4.61×10^{-3}

22. 1.218×10^{-8}

23. 2.0×10^{-3}

24. 7.7521×10^{-5}

25. 9.0×10^7

26. 2.8×10^2

27. 6700

28. 39,920,000

29. 0.0003221

30. 0.57

31. 0.0000721

32. 0.001947

33. 607,400

34. 169.98

35. 5.088×10^{12}

36. 4.63×10^{-2}

37. 5.1641×10^{10}

38. 5.4×10^{-7}

39. 7,500,000

40. 0.0525

41. \$55,000,000,000

42. 0.000001539

43. 1.188×10^{12}

44. 1.84×10^4

45. 1.37×10^{-41}

Section 5.2

1. $(2^2)^3 = 2^{2 \cdot 3} = 2^6 = 64$

2. $\dfrac{1}{16}$

3. $(z^2)^5 = z^{2 \cdot 5} = z^{10}$

4. p^{12}

5. $(3q^6)^2 = 3^2 (q^6)^2 = 9q^{6 \cdot 2} = 9q^{12}$

6. $\dfrac{w^{10}}{4}$

7. $(3xy^7)^3 = 3^3 x^3 y^{7 \cdot 3} = 27x^3 y^{21}$

8. $\dfrac{b^{24} c^{42}}{a^{18}}$

9. $\dfrac{3x^2 y^{-4} z^3}{(xyz)^{-3}} = \dfrac{3x^2 y^{-4} z^3}{x^{-3} y^{-3} z^{-3}} = \dfrac{3x^5 z^6}{y}$

10. $\dfrac{y^2}{256xz^2}$

11. $(3 \times 10^2)(1.7 \times 10^7) = 3 \times 10^2 \times 1.7 \times 10^7$
$$= 5.1 \times 10^9$$

12. 2.31×10^{12}

13. $(5.9 \times 10^{-5})(7.5 \times 10^6) = 5.9 \times 10^{-5} \times 7.5 \times 10^6$
$$= 44.25 \times 10^1$$
$$= (4.425 \times 10^1) \times 10^1$$
$$\approx 4.43 \times 10^2$$

14. 3.21×10^{-5}

15. $(9 \times 10^{-2})^2 = 9^2 \times (10^{-2})^2$
$$= 81 \times 10^{-4}$$
$$= (8.1 \times 10^1) \times 10^{-4}$$
$$= 8.1 \times 10^{-3}$$

16. 6.4×10^{25}

17. $\dfrac{8.7 \times 10^6}{4.9 \times 10^7} = \left(\dfrac{8.7}{4.9}\right)\left(\dfrac{10^6}{10^7}\right) = 1.78 \times 10^{-1}$

18. 2.65×10^2

19. $\dfrac{1.02 \times 10^{-4}}{3.01 \times 10^{-3}} = \left(\dfrac{1.02}{3.01}\right)\left(\dfrac{10^{-4}}{10^{-3}}\right)$
$$= (3.39 \times 10^{-1}) \times (10^{-1})$$
$$= 3.39 \times 10^{-2}$$

20. 2.20×10^9

21. $\dfrac{(9 \times 10^{-3})(5.7 \times 10^4)}{3.9 \times 10^7} = \dfrac{9 \times 5.7}{3.9} \times 10^{-3+4-7}$
$$= (1.32 \times 10^1) \times 10^{-6}$$
$$= 1.32 \times 10^{-5}$$

22. 2.72×10^4

23. $\dfrac{0.0000018}{0.0003} = \dfrac{1.8 \times 10^{-6}}{3.0 \times 10^{-4}}$
$$= (6.0 \times 10^{-1}) \times (10^{-2})$$
$$= 6.0 \times 10^{-3}$$

24. 3.0×10^{-9}

25. $\dfrac{0.007 \times 1000}{0.0005} = \dfrac{7.0 \times 10^{-3} \times 10^3}{5.0 \times 10^{-4}}$
$$= 1.4 \times 10^{-3+3+4}$$
$$= 1.4 \times 10^4$$

26. 9.0×10^{-2}

27. $D = \dfrac{M}{V} = \dfrac{88,000}{0.11}$
$$= \dfrac{8.8 \times 10^4}{1.1 \times 10^{-1}}$$
$$= 8.0 \times 10^{4-(-1)}$$
$$= 8.0 \times 10^5 \text{ pounds per cubic foot}$$

28. 2.375×10^3 pounds per cubic foot

29. $\dfrac{3.36 \times 10^{12}}{2.49 \times 10^8} = \$1.35 \times 10^4$ per person

30. $\$1.85 \times 10^4$ per person

Section 5.3

1. $7x^6$ is of degree 6.
$-2x^9$ is of degree 9.
The polynomial is of degree 9.

2. $3x^4$ is of degree 4.
$-x^3$ is of degree 3.
$10x^2$ is of degree 2.
12 is of degree 0.
The polynomial is of degree 4.

3. 1 is of degree 0.
$-y^2$ is of degree 2.
y^5 is of degree 5.
The polynomial is of degree 5.

4. $2x^2$ is of degree 2.
-1 is of degree 0.
The polynomial is of degree 2.

5. $P(0) = 3(0)^2 - 5(0) + 20$
$\qquad = 20$

6. 32

7. $P(-1) = 3(-1)^2 - 5(-1) + 20$
$\qquad = 28$

8. -42

9. $Q(3) = 5(3)^3 - 2$
$\qquad = 133$

10. -2

11. $P(82) = -0.22(82)^2 + 45.7(82) - 2031.3$
$\qquad = 236.82 \approx 237$
237 people attend.

12. 1999 feet, 1807 feet

13. $R(60,000) = 50 + 0.3(60,000)$
$\qquad = \$18,050$

14. 2079 feet, 1967 feet, 1727 feet

15. $(5x^2 - 3) + (2x^2 + 7) = 5x^2 + 2x^2 - 3 + 7$
$\qquad = 7x^2 + 4$

16. $2y^3 + 7y + 2$

17. $2(x + 7) + (x^2 - 5x + 3) = 2x + 14 + x^2 - 5x + 3$
$\qquad = x^2 - 3x + 17$

18. $6x^3$

19. $4w^3 - 2w^2 + 10w - 1 - 2w^3 + 6w^2 - 5w - 7$
$\qquad = 4w^3 - 2w^3 - 2w^2 + 6w^2 + 10w - 5w - 1 - 7$
$\qquad = 2w^3 + 4w^2 + 5w - 8$

20. $2x + 21$

21. $(2x^2 + x - 6) - (x^2 - 5x + 17)$
$\qquad = 2x^2 + x - 6 - x^2 + 5x - 17$
$\qquad = 2x^2 - x^2 + x + 5x - 6 - 17$
$\qquad = x^2 + 6x - 23$

22. $3x^3 - 12x + 13$

23. $(19 - 3y - 7y^2) - (y^3 - y^2 + 11y - 1)$
$\qquad = 19 - 3y - 7y^2 - y^3 + y^2 - 11y + 1$
$\qquad = -y^3 - 7y^2 + y^2 - 3y - 11y + 19 + 1$
$\qquad = -y^3 - 6y^2 - 14y + 20$

24. $w^2 + 3w + 2$

25. $9.3x^2 - 8.5x - 2.9$

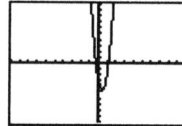

26. $8x^3 + 1.3x^2 - 8.7x + 9.7$

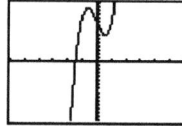

Section 5.4

1. $(7x^6)(-2x^9) = -14x^{15}$

2. $30x^6 y^3$

3. $(12st)(-2st) = -24s^2 t^2$

4. $-2x^3 y$

5. $5t(t - 3) = (5t)t - (5t)3$
$\qquad = 5t^2 - 15t$

6. $12z - 12z^2$

7. $-2x(ax + 2x^2 - 4) = (-2x)ax + (-2x)2x^2 - (-2x)4$
$\qquad = -2ax^2 - 4x^3 + 8x$

8. $3b - 2bx$

9. $(x + 2)(x - 6) = x \cdot x + (-6)x + 2x - 12$
$\qquad = x^2 - 6x + 2x - 12$
$\qquad = x^2 - 4x - 12$

10. $d^2 - 6d - 7$

11. $(3x + 1)(4x + 6) = 3x \cdot 4x + 3x(6) + 1 \cdot 4x + 1 \cdot 6$
$\qquad = 12x^2 + 18x + 4x + 6$
$\qquad = 12x^2 + 22x + 6$

12. $\dfrac{1}{4}m^2 - 4m - 20$

13. $(2y + 4x)(5y - x)$

$= 2y \cdot 5y + 2y(-x) + 4x \cdot 5y + 4x(-x)$

$= 10y^2 - 2xy + 20xy - 4x^2$

$= 10y^2 + 18xy - 4x^2$

14. $48r^2 - 38r + 4$

15. $(t - 3)^2 = t^2 - 2(3)t + 3^2$

$\qquad\qquad = t^2 - 6t + 9$

16. $w^2 + 12w + 36$

17. $(4v - 2)(4v + 2) = (4v)^2 - 2^2$

$\qquad\qquad\qquad = 16v^2 - 4$

18. $4q^2 - 81$

19. $(2z + 5)^2 = (2z)^2 + 2(5)(2z) + 5^2$

$\qquad\qquad = 4z^2 + 20z + 25$

20. $16y^2 - 80y + 100$

21. $(3a + 8b)(3a - 8b) = (3a)^2 - (8b)^2$

$\qquad\qquad\qquad = 9a^2 - 64b^2$

22. $4x^2 - y^2$

23. $Q(x) \cdot R(x) = (-4x)(3x - 1)$

$\qquad\qquad = (-4x)(3x) - (-4x)(1)$

$\qquad\qquad = -12x^2 + 4x$

24. $-4x^3 - 8x^2 - 4x$

25. $R(x) \cdot P(x) = (3x - 1)(x^2 + 2x + 1)$

$\qquad\qquad = 3x^3 + 6x^2 + 3x - x^2 - 2x - 1$

$\qquad\qquad = 3x^3 + 5x^2 + x - 1$

26. $9x^2 - 6x + 1$

27. $Volume = 2\pi r h$

$\qquad = 2\pi(x + 2)(15)$

$\qquad = 30\pi(x + 2)$

$\qquad = 30\pi x + 60\pi \quad \text{cubic cm}$

28. $144\pi r^2 - 216\pi r + 81\pi \quad \text{square meters}$

29. $36x^2 - 12x + 1$

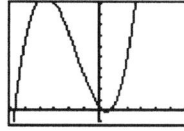

30. $4x^2 - 25$

31. $3x^3 + 14x^2 - 17x + 4$

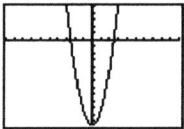

Section 5.5

1. $2x^3$

2. x

3. $2st$

4. $5q^2r^3$

5. $t(5t - 3)$

6. $12z(1 - z)$

7. $2x(-ax - 2ax^2 + 4)$

8. $4b(3 - 2x)$

9. $(x + 2)(5b + 3)$

10. $(d - 7)(4x - y)$

11. $(3x + 1)(4x + 5)$

12. $3w^2y(4y^2 - 1 + 3y)$

13. $xy + 2x + y + 2 = (xy + 2x) + (y + 2)$

$\qquad\qquad = x(y + 2) + 1(y + 2)$

$\qquad\qquad = (y + 2)(x + 1)$

14. $(a - 3)(b + 3)$

15. $x^2 - 2x + xy - 2y = (x^2 - 2x) + (xy - 2y)$

$\qquad\qquad = x(x - 2) + y(x - 2)$

$\qquad\qquad = (x - 2)(x + y)$

16. $(a + b)(a - 12)$

17. $21xy - 35x + 6y - 10 = (21xy - 35x) + (6y - 10)$

$\qquad\qquad = 7x(3y - 5) + 2(3y - 5)$

$\qquad\qquad = (3y - 5)(7x + 2)$

18. $(3y - 1)(2x - 1)$

19. $2x^2 + x - 2xy - y = (2x^2 + x) + (-2xy - y)$

$\qquad\qquad = x(2x + 1) - y(2x + 1)$

$\qquad\qquad = (2x + 1)(x - y)$

20. $(a - 2)(a + 2b)$

21. $\frac{1}{2}ah + \frac{1}{2}bh = \frac{1}{2}h(a + b)$

22. $x(2y - 1)(x + 2)$

23. $(x - 2)(3x^2 + 5)$

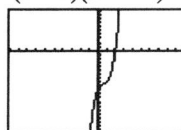

24. $(x + 3)(x^3 - 3)$

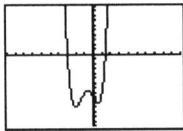

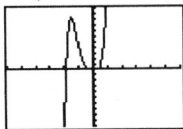

Section 5.6

1. $(x-2)(x-1)$
2. $(x-5)(x+1)$
3. $(x+5)(x+3)$
4. $(x+7)(x-3)$
5. $2x^2 + 36x + 160 = 2(x^2 + 18x + 80)$
   $$= 2(x+10)(x+8)$$
6. $3(x-7)(x+6)$
7. $(2x+1)(x+1)$
8. $(2x+1)(x-9)$
9. $(3x+2)(2x-5)$
10. $(5x-3)(2x+1)$
11. $8x^2 - 42x - 36 = 2(4x^2 - 21x - 18)$
    $$= 2(4x+3)(x-6)$$
12. $3y(3x-1)(3x+2)$
13. $(x-7)(x-9)$
14. $5s(t-7)(t-9)$
15. $(2x-3)(x+9)$
16. $3w(2w-3)(w+9)$
17. $2(3x-1)^2 + 9(3x-1) - 110$
    $$= [2(3x-1)-11][(3x-1)+10]$$
    $$= (6x-2-11)(3x-1+10)$$
    $$= (6x-13)(3x+9)$$
    $$= 3(6x-13)(x+3)$$
18. $(y^2+3)(y^2+10)$
19. $q(q^2-4)(q^2-1)$
20. $6x(x^2+3)(x^2-2)$
21. (a) $V(x) = 9x^3 - 18x^2 + 8x$
    $$= x(9x^2 - 18x + 8)$$
    $$= x(3x-2)(3x-4)$$
    (b) $3(7) - 2 = 21 - 2 = 19$
    $$3(7) - 4 = 21 - 4 = 17$$
    $Volume = 7(19)(17) = 2261$ cm^3
22. Not completely factored.
    Should be: $2(x+4)(x-3)$
23. $x^3(5x^2 + 10x + 1)$

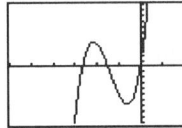

24. $x(2x+3)(3x+8)$

Section 5.7

1. $(x-4)^2$
2. $(3x+8)^2$
3. $(2x-5)^2$
4. $2(z-10)^2$
5. $5(y-3)^2$
6. $3(2x+7)^2$
7. $(4w+3z)(4w-3z)$
8. $2z(5x+3y)(5x-3y)$
9. $(6r+13q)(6r-13q)$
10. $3(x+2yz)(x-2yz)$
11. $(w-10)(w^2+10w+100)$
12. $(x+8)(x^2-8x+64)$
13. $(2y+3x)(4y^2-6xy+9x^2)$
14. $r^2(4m-5n)(16m^2+20mn+25n^2)$
15. $z^6 - 64$
    $$(z^3)^2 - (8)^2$$
    $$(z^3+8)(z^3-8)$$
    $$(z+2)(z^2-2z+4)(z-2)(z^2+2z+4)$$
16. $\left(\dfrac{1}{5} - 3w\right)\left(\dfrac{1}{5} + 3w\right)$
17. $\left(4y - \dfrac{3}{4}x\right)^2$
18. $9(2w+1)(4w^2+2w+1)$
19. $7\left(3x + \dfrac{1}{2}\right)^2$
20. $3b(a+6b)(a-6b)$
21. $3(5x-3)^2$

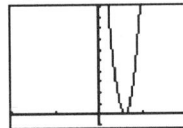

22. $2x(6-7x)(6+7x)$

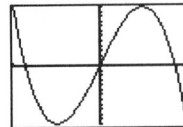

Section 5.8

1. $(d-7)(d+1) = 0$

$d - 7 = 0$

$d = 7$

$d + 1 = 0$

$d = -1$

$\{-1, 7\}$

2. $\left\{-\dfrac{1}{3}, -\dfrac{3}{2}\right\}$

3. $x^2 - 4x - 5 = 0$

$(x - 5)(x + 1) = 0$

$x = 5$

$x = -1$

$\{-1, 5\}$

4. $\{4\}$

5. $x^2 + 4x - 18 = 3$

$x^2 + 4x - 21 = 0$

$(x + 7)(x - 3) = 0$

$x = -7$

$x = 3$

$\{-7, 3\}$

6. $\{-6, 7\}$

7. $2x(4x - 21) + 10 = 46$

$8x^2 - 42x + 10 - 46 = 0$

$8x^2 - 42x - 36 = 0$

$2(4x^2 - 21x - 18) = 0$

$2(4x + 3)(x - 6) = 0$

$x = -\dfrac{3}{4}$

$x = 6$

$\left\{-\dfrac{3}{4}, 6\right\}$

8. $\left\{-\dfrac{8}{3}\right\}$

9. $2(3x - 1)^2 + 9(3x - 1) - 110 = 0$

$[2(3x - 1) - 11][(3x - 1) + 10] = 0$

$2(3x - 1) - 11 = 0$

$6x - 2 - 11 = 0$

$6x = 13$

$x = \dfrac{13}{6}$

$3x - 1 + 10 = 0$

$3x = -9$

$x = -3$

$\left\{-3, \dfrac{13}{6}\right\}$

10. $\left\{-\dfrac{7}{2}\right\}$

11. $x(x + 2) = 288$

$x^2 + 2x - 288 = 0$

$(x + 18)(x - 16) = 0$

$x = -18$

$x = 16$

The two numbers are 16 and 18, or −16 and −18.

12. 7 and 9 or −7 and −9

13. $\sqrt{100^2 + 240^2} = 260$

$100 + 240 = 340$

$340 - 260 = 80$

You save 80 yards.

14. 7 seconds

15. (0.31, 0), (3.19, 0)

16. (−4.46, 0), (1.12, 0)

17. Approximately 5.89 seconds

Section 5.9

1. $x = \dfrac{-8}{2(1)} = -4$

$f(-4) = (-4)^2 + 8(-4) + 15$

$= -1$

Vertex: (−4, −1)

2. Vertex: (−9, −2)

3. $x = \dfrac{-(-3)}{2(3)} = \dfrac{1}{2}$

$f\left(\dfrac{1}{2}\right) = 3\left(\dfrac{1}{2}\right)^2 - 3\left(\dfrac{1}{2}\right) - 126$

$= -126.75$

Vertex: $\left(\dfrac{1}{2}, -126.75\right)$

4. (2.625, −91.125)

5. $f(0) = 0^2 + 4(0) - 21 = -21$

$0 = x^2 + 4x - 21$

$0 = (x + 7)(x - 3)$

$x = -7$

$x = 3$

(0, −21), (−7, 0), (3, 0)

6. (−3.5, 0), (0, 147)

7.

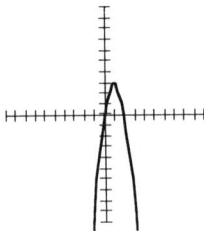

8.

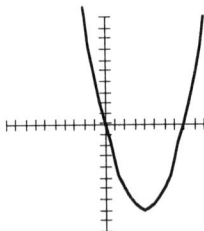

9.

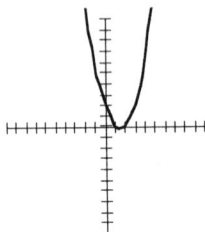

10.

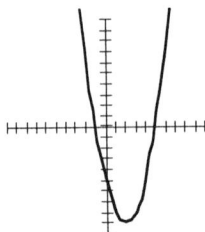

11.

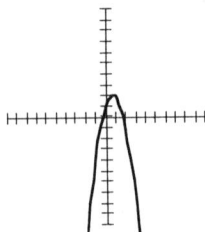

12.

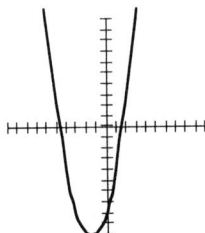

13.

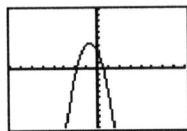

$(-1.87, 0)$, $(0.54, 0)$

14.

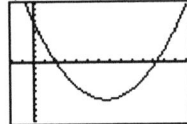

$(2.07, 0)$, $(11.26, 0)$

Chapter 5 Practice Test

1. $\dfrac{1}{27y^3}$

2. $-\dfrac{1}{2}q$

3. $\dfrac{m^5}{3n^3}$

4. $\dfrac{2a^6b}{c}$

5. 5.673×10^7

6. 3.54×10^{-4}

7. 0.000823

8. $619,000,000$

9. $2x^2 - 22x + 11$

10. $-12a^3b - 6ab^2 + 18a^2b^2$

11. $6w^2 + 11wz - 35z^2$

12. $4x^2 - 20x + 25$

13. $9x^3 - 100$

14. $6z^3 - 2z^2 - 5z + 1$

15. $5xy^3(2x^2 - y)$

16. $(x + 3)(x + 11)$

17. $(3y - 1)^2$

18. $4x(2x + 1)(x - 1)$

19. $(3x + 13)(3x - 13)$

20. $(x - 5)(x^2 + 5x + 25)$

21. $7(2x^2 + 9)$

22. $(x + 7)(x + 2)(x - 2)$

23. $\left\{-9, \dfrac{7}{2}\right\}$

24. $\{-1, 1\}$

25.
$$9x^3 - 18x^2 - x + 4 = 2$$
$$9x^3 - 18x^2 - x + 2 = 0$$
$$9x^2(x-2) - 1(x-2) = 0$$
$$(x-2)(9x^2-1) = 0$$
$$(x-2)(3x+1)(3x-1) = 0$$
$$x = 2$$
$$x = -\frac{1}{3}$$
$$x = \frac{1}{3}$$
$$\left\{ -\frac{1}{3}, \frac{1}{3}, 2 \right\}$$

26. $\{-6, 6\}$

27.

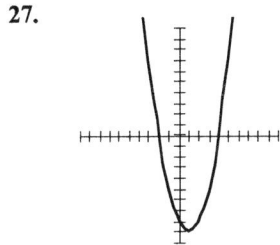

28.

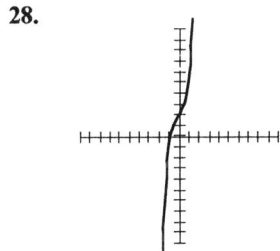

Chapter 6

Section 6.1

1.
$$f(0) = \frac{0^2 + 3 \cdot 0 - 10}{5 \cdot 0 + 4} = \frac{-10}{4} = -\frac{5}{2}$$
$$f(4) = \frac{4^2 + 3 \cdot 4 - 10}{5 \cdot 4 + 4} = \frac{18}{24} = \frac{3}{4}$$

2. $\dfrac{11}{9}, -\dfrac{103}{3}$

3. $\{q \mid q \text{ is a real number}\}$

4. $\{t \mid t \text{ is a real number and } t \neq -7\}$

5. $\left\{ x \mid x \text{ is a real number and } x \neq -\dfrac{3}{2} \right\}$

6. $\{y \mid y \text{ is a real number and } y \neq 5 \text{ and } y \neq 1\}$

7.
$$\frac{4r^2 + 20r}{4r^3 - 8r^2 + 8r} = \frac{4r(r+5)}{4r(r^2 - 2r + 2)} = \frac{r+5}{r^2 - 2r + 2}$$

8. -1

9.
$$\frac{y^2 + 4y + 3}{2y^2 - y - 3} = \frac{(y+3)(y+1)}{(2y-3)(y+1)} = \frac{y+3}{2y-3}$$

10. $2y + 7$

11.
$$\frac{3}{7y^2} = \frac{?}{7y^3 + 7y^2} = \frac{3(y+1)}{7y^2(y+1)}$$

12.
$$\frac{2}{y} = \frac{?}{3y^2 - 5y} = \frac{2(3y-5)}{y(3y-5)}$$

13.
$$\frac{x}{x+3} = \frac{?}{x^2 - 6x - 27} = \frac{x(x-9)}{(x+3)(x-9)}$$

14.
$$\frac{q-3}{3q+1} = \frac{?}{9q^2 - 1} = \frac{(q-3)(3q-1)}{(3q+1)(3q-1)}$$

15.

(a) $R(1) = \dfrac{80(1)^2}{2(1)^2 - 3(1) + 4} = \dfrac{80}{3}$

Revenue in the first year is approximately $2,666,667.

(b) $R(2) = \dfrac{80(2)^2}{2(2)^2 - 3(2) + 4} = \dfrac{160}{3}$

Revenue in the second year is approximately $5,333,333.

(c) $R(3) = \dfrac{80(3)^2}{2(3)^2 - 3(3) + 4} = \dfrac{720}{13}$

Revenue in the third year is approximately $5,538,462.

(d) $R(4) = \dfrac{80(4)^2}{2(4)^2 - 3(4) + 4} = \dfrac{160}{3}$

Revenue in the fourth year is approximately $5,333,333.

16. $\left\{ x \mid x \text{ is a real number and } x \neq \dfrac{1}{2} \text{ and } x \neq 3 \right\}$

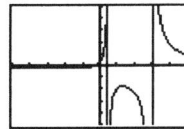

17.
$$f(x) = \frac{4x^2 + 5}{6x^2 + 33x + 27} = \frac{4x^2 + 5}{3(2x+9)(x+1)}$$

$\left\{ x \mid x \text{ is a real number and } x \neq -\dfrac{9}{2} \text{ and } x \neq -1 \right\}$

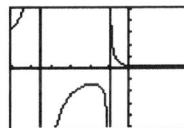

Section 6.2

1. $\dfrac{2y-10}{y^2+3y}\cdot\dfrac{y^2}{5-y}=\dfrac{-2(5-y)}{y(y+3)}\cdot\dfrac{y^2}{5-y}=\dfrac{-2y}{y+3}$

2. $\dfrac{2y+1}{2(y+3)}$

3. $\dfrac{3a+6}{a^2-1}\div\dfrac{a^2+4a+4}{4a-4}=\dfrac{3a+6}{a^2-1}\cdot\dfrac{4a-4}{a^2+4a+4}$

$\qquad\qquad=\dfrac{3(a+2)}{(a+1)(a-1)}\cdot\dfrac{4(a-1)}{(a+2)(a+2)}$

$\qquad\qquad=\dfrac{12}{(a+1)(a+2)}$

4. $\dfrac{5}{2x(x-2)}$

5. $\dfrac{4y^2+4y-15}{4y^2+20y+25}\cdot\dfrac{y^2-4y-5}{3-2y}$

$\qquad=\dfrac{(2y+5)(2y-3)}{(2y+5)(2y+5)}\cdot\dfrac{(y+1)(y-5)}{-1(2y-3)}$

$\qquad=-\dfrac{(y+1)(y-5)}{2y+5}$

6. $\dfrac{(x^2+x+2)(x+2)}{x-1}$

7. $f(x)+g(x)=9-3x+x^2+x-12$

$\qquad\qquad=x^2-2x-3$

8. $x^2+4x-21$

9. $f(x)g(x)=(9-3x)(x^2+x-12)$

$\qquad\qquad=9x^2+9x-108-3x^3-3x^2+36x$

$\qquad\qquad=-3x^3+6x^2+45x-108$

10. $-\dfrac{x+4}{3}$

11. $(g-f)(2)=2^2+4(2)-21$

$\qquad\qquad=-9$

12. -108

13. $\text{Area }=\dfrac{1}{2}bh$

$\qquad=\dfrac{1}{2}\left(\dfrac{8x^2}{x+2}\right)\left(\dfrac{1}{3x}\right)=\dfrac{4x}{3(x+2)}$

14. $y(y-1)$ square inches

Section 6.3

1. $\dfrac{10}{5y-y^2}-\dfrac{y+5}{5y-y^2}=\dfrac{10-y-5}{y(5-y)}=\dfrac{5-y}{y(5-y)}=\dfrac{1}{y}$

2. $\dfrac{1}{y+9}$

3. $\dfrac{3}{x+5},\dfrac{2x}{5-x}\quad\Rightarrow\quad\dfrac{3}{x+5},\dfrac{2x}{-(x-5)}$

LCD: $-(x+5)(x-5)$

4. $12xy^2z$

5. $\dfrac{t}{t^2-16},\dfrac{11}{t+4}\quad\Rightarrow\quad\dfrac{t}{(t+4)(t-4)},\dfrac{11}{t+4}$

LCD: $(t+4)(t-4)$

6. $z(z-6)(z-1)$

7. $\dfrac{x}{x+3}+\dfrac{4}{x+4}=\dfrac{x(x+4)}{(x+3)(x+4)}+\dfrac{4(x+3)}{(x+4)(x+3)}$

$\qquad\qquad=\dfrac{x^2+4x+4x+12}{(x+3)(x+4)}$

$\qquad\qquad=\dfrac{x^2+8x+12}{(x+3)(x+4)}$

8. $\dfrac{-14x^2+32x-9}{(2x+3)(2x-3)}$

9. $\dfrac{10}{x}-\dfrac{2}{x-1}=\dfrac{10(x-1)-2x}{x(x-1)}$

$\qquad\qquad=\dfrac{10x-10-2x}{x(x-1)}$

$\qquad\qquad=\dfrac{8x-10}{x(x-1)}$

10. $\dfrac{x^2+11}{(3x+1)(x-2)(x+3)}$

11. $\dfrac{2x}{x-1}+\dfrac{3}{x+1}-\dfrac{2x}{x^2-1}=\dfrac{2x(x+1)+3(x-1)-2x}{(x+1)(x-1)}$

$\qquad\qquad=\dfrac{2x^2+2x+3x-3-2x}{(x+1)(x-1)}$

$\qquad\qquad=\dfrac{2x^2+3x-3}{(x+1)(x-1)}$

12. $\dfrac{-3x^2+17x+22}{(x-2)^2(x+2)}$

13. $\dfrac{-x^2-10x-7}{2(3x-1)(x+2)(x-1)}$

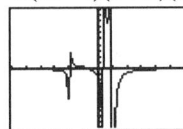

14. $\dfrac{5(2x-3)}{-6(x+2)}$

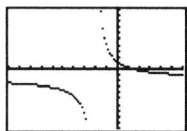

Section 6.4

1. $\dfrac{\dfrac{3x}{2}}{\dfrac{x^2}{4}} = \dfrac{3x}{2} \cdot \dfrac{4}{x^2} = \dfrac{6}{x}$

2. $\dfrac{2}{3}$

3. $\dfrac{2 + \dfrac{3x}{2}}{4 - \dfrac{5x}{3}} = \dfrac{6 \cdot 2 + 6 \cdot \dfrac{3x}{2}}{6 \cdot 4 - 6 \cdot \dfrac{5x}{3}} = \dfrac{12 + 9x}{24 - 10x} = \dfrac{3(4 + 3x)}{2(12 - 5x)}$

4. $\dfrac{4x - 5}{8x + 12}$

5. $\dfrac{\dfrac{5x}{4x^2 - 9}}{\dfrac{7x^2 - 3x}{2x + 3}} = \dfrac{5x}{4x^2 - 9} \cdot \dfrac{2x + 3}{7x^2 - 3x}$

$= \dfrac{5x}{(2x - 3)(2x + 3)} \cdot \dfrac{2x + 3}{x(7x - 3)}$

$= \dfrac{5}{(2x - 3)(7x - 3)}$

6. $\dfrac{w - 3}{w + 3}$

7. $\dfrac{y^{-2}}{y^{-1} + (2y)^{-1}} = \dfrac{\dfrac{1}{y^2}}{\dfrac{1}{y} + \dfrac{1}{2y}} = \dfrac{2}{2y + y} = \dfrac{2}{3y}$

8. $\dfrac{x^4}{x + 1}$

9. $\dfrac{5 - w^{-2}}{1 + w^{-1}} = \dfrac{5 - \dfrac{1}{w^2}}{1 + \dfrac{1}{w}} = \dfrac{5w^2 - 1}{w^2 + w}$

10. $\dfrac{3r^2 + 4q^2 r}{2qr^2 + q^2}$

11.

(a) $f(a + h) = \dfrac{1}{a + h - 2}$

(b) $f(a) = \dfrac{1}{a - 2}$

(c) $\dfrac{f(a + h) - f(a)}{h} = \dfrac{\dfrac{1}{a + h - 2} - \dfrac{1}{a - 2}}{h}$

(d) $\dfrac{\dfrac{1}{a + h - 2} - \dfrac{1}{a - 2}}{h} = \dfrac{a - 2 - (a + h - 2)}{h(a - 2)(a + h - 2)}$

$= \dfrac{a - 2 - a - h + 2}{h(a - 2)(a + h - 2)}$

$= \dfrac{-h}{h(a - 2)(a + h - 2)}$

$= \dfrac{-1}{(a - 2)(a + h - 2)}$

12.

(a) $\dfrac{a + h}{a + h + 3}$

(b) $\dfrac{a}{a + 3}$

(c) $\dfrac{\dfrac{a + h}{a + h + 3} - \dfrac{a}{a + 3}}{h}$

(d) $\dfrac{3}{(a + 3)(a + h + 3)}$

13. $\dfrac{3(2x - 1)}{x + 3}$

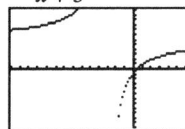

14. $\dfrac{3 - x}{2x + 5x^2}$

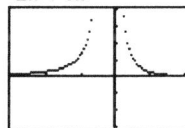

15. 39.6%
16. 29.2%
17. Midsize, 6%
18. Small, 9.9%
19. Increased by 3.8%

Section 6.5

1. $\dfrac{90x^2 + 50x - 10}{10x} = \dfrac{90x^2}{10x} + \dfrac{50x}{10x} - \dfrac{10}{10x} = 9x + 5 - \dfrac{1}{x}$

2. $5 + \dfrac{3}{a} - \dfrac{9}{a^2}$

3. $\dfrac{56q^2 r^2 - 24q^2 r + 28qr^2 + 4qr}{4q^2 r^2}$

$= 16 - \dfrac{6}{r} + \dfrac{7}{q} + \dfrac{1}{qr}$

4. $\dfrac{2}{y} + \dfrac{5x}{3y^2} - \dfrac{3}{y^2}$

5.
$$y+6\overline{\smash{\big)}\,2y^2+9y-18} \quad \frac{2y-3}{}$$
$$\underline{2y^2+12y}$$
$$-3y-18$$
$$\underline{-3y-18}$$

$$\frac{2y^2+9y-18}{y+6}=2y-3$$

6. $x+3$

7.
$$3w^2+0w-1\overline{\smash{\big)}\,3w^3+12w^2-w-2} \quad \frac{w+4}{}$$
$$\underline{3w^3+0w^2-w}$$
$$12w^2-0w-2$$
$$\underline{12w^2-0w-4}$$
$$2$$

$$w+4+\frac{2}{3w^2-1}$$

8. $2x^3+1+\dfrac{6}{3x-1}$

9. $\dfrac{15x^3-3x^2+40x+85}{5}=3x^3-\dfrac{3}{5}x^2+8x+17$

meters

10. $2x-1$

11.
$$2x-1\overline{\smash{\big)}\,4x^2+12x-7} \quad \frac{2x+7}{}$$
$$\underline{4x^2-2x}$$
$$14x-7$$
$$\underline{14x-7}$$

The walker walks at a speed of $2x+7$ mph.

12. $2x-3$ cm

13. $x-5$

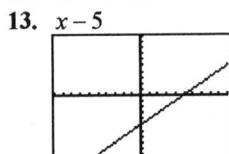

14. $x+2-\dfrac{3}{x+3}$

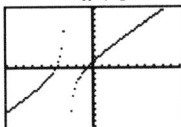

Section 6.6

1.
$$\begin{array}{r|rrr} 1 & 9 & 5 & -7 \\ & & 9 & 14 \\ \hline & 9 & 14 & 7 \end{array}$$
$$9x+14+\frac{7}{x-1}$$

2. $a^2-10a-50-\dfrac{350}{a-5}$

3.
$$\begin{array}{r|rrr} -5 & 4 & 19 & -5 \\ & & -20 & 5 \\ \hline & 4 & -1 & 0 \end{array}$$
$$4q-1$$

4. $2y-3$

5.
$$\begin{array}{r|rrrrr} 1 & 6 & -2 & 0 & 3 & 5 \\ & & 6 & 4 & 4 & 7 \\ \hline & 6 & 4 & 4 & 7 & 12 \end{array}$$
$$6x^3+4x^2+4x+7+\frac{12}{x-1}$$

6. $2x^4+x^3+3x^2+6x+12+\dfrac{26}{x-3}$

7.
$$\begin{array}{r|rrrr} -4 & 3 & 12 & -1 & -2 \\ & & -12 & 0 & 4 \\ \hline & 3 & 0 & -1 & 2 \end{array}$$
$$P(-4)=2$$

8. 4921

9.
$$\begin{array}{r|rrrrrr} -5 & 1 & 1 & 0 & 0 & 0 & 500 \\ & & -5 & 20 & -100 & 500 & -2500 \\ \hline & 1 & -4 & 20 & -100 & 500 & -2000 \end{array}$$
$$P(-5)=-2000$$

10. 16

11. $\dfrac{15y^3+354y^2-144y}{3y}=5y^2+118y-48$
$$\begin{array}{r|rrr} -24 & 5 & 118 & -48 \\ & & -120 & 48 \\ \hline & 5 & -2 & 0 \end{array}$$

The height is $5y-2$ centimeters.

12. $2x^2+3x-21$ feet

Section 6.7

1.
$$\frac{a}{4}-\frac{a}{6}=\frac{2}{3}$$
$$12\left(\frac{a}{4}-\frac{a}{6}\right)=12\left(\frac{2}{3}\right)$$
$$3a-2a=8$$
$$a=8$$
The solution set is $\{8\}$.

2. $\{-11\}$

3.
$$\frac{2x-7}{6x+5} = \frac{x-4}{3x-1}$$
$$(2x-7)(3x-1) = (x-4)(6x+5)$$
$$6x^2 - 23x + 7 = 6x^2 - 19x - 20$$
$$-4x = -27$$
$$x = \frac{27}{4}$$

The solution set is $\left\{\frac{27}{4}\right\}$.

4. $\{1\}$

5.
$$\frac{21}{(y+3)^2} = 2 - \frac{1}{y+3}$$
$$21 = 2(y+3)^2 - (y+3)$$
$$21 = 2(y^2 + 6y + 9) - y - 3$$
$$21 = 2y^2 + 12y + 18 - y - 3$$
$$0 = 2y^2 + 11y - 6$$
$$0 = (2y - 1)(y + 6)$$
$$y = \frac{1}{2} \quad \text{or} \quad y = -6$$

The solution set is $\left\{-6, \frac{1}{2}\right\}$.

6. $\left\{-\frac{1}{11}\right\}$

7.
$$\frac{3x}{x-3} - \frac{18}{x^2 - 4x + 3} = \frac{11}{x-1}$$
$$3x(x-1) - 18 = 11(x-3)$$
$$3x^2 - 3x - 18 = 11x - 33$$
$$3x^2 - 14x + 15 = 0$$
$$(3x - 5)(x - 3) = 0$$
$$x = \frac{5}{3} \quad \text{or} \quad x = 3, \text{ extraneous}$$

The solution set is $\left\{\frac{5}{3}\right\}$.

8. $\{ \ \}$

9.
$$20 = 17.50 + \frac{5000}{x}$$
$$20x = 17.50x + 5000$$
$$2.5x = 5000$$
$$x = 2000$$

2000 calculators must be produced for the average cost to be $20.

10. 800 boxes

11.
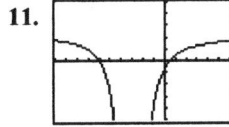

12.
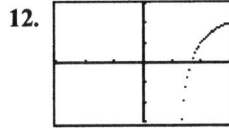

13.
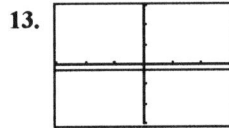

14.
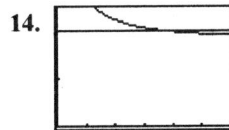

Section 6.8

1.
$$x + \frac{9}{x} = 6$$
$$x^2 + 9 = 6x$$
$$x^2 - 6x + 9 = 0$$
$$(x-3)(x-3) = 0$$
$$x = 3$$

The number is 3.

2.   The numbers are 7 and -3.

3.
$$\frac{x}{\frac{2}{x}} = 8$$
$$x \cdot \frac{x}{2} = 8$$
$$x^2 = 16$$
$$x = \pm 4$$

The numbers are 4 and -4.

4.   The numbers are $\frac{3}{5}$ and $-\frac{3}{5}$.

5.
$$\frac{12}{r-8} = \frac{24}{r+8}$$
$$12(r+8) = 24(r-8)$$
$$12r + 96 = 24r - 192$$
$$288 = 12r$$
$$24 = r$$

The speed of the boat in still water is 24 mph.

6.   The speed of the skater is $6\frac{2}{3}$ mph.

7. $\dfrac{1}{6} - \dfrac{1}{12} = \dfrac{1}{t}$

$2t - t = 12$

$t = 12$

It will take 12 hours to empty the warehouse.

8. It will take 12 hours for the second person to rake the yard working alone.

9. $\dfrac{1}{6} - \dfrac{1}{36} = \dfrac{1}{t}$

$6t - t = 36$

$5t = 36$

$t = 7.2$

It will take 7.2 hours to fill the pool.

10. It will take 42 hours to empty the vat.

Section 6.9

1. $P = \dfrac{k}{q}$

2. $W = kz$

3. $G = kts$

4. $B = kxy^2$

5. $I = \dfrac{V}{R}$

$32 = \dfrac{V}{25}$

$800 = V$

$I = \dfrac{800}{20} = 40$

The current is 40 amperes when the resistance is 20 ohms.

6. Approximately 2813 ice cream cones will be sold when the temperature is 75°.

7. $h = ksc$

$72 = k(32)(3)$

$0.75 = k$

$h = 0.75(21)(4) = 63$

The professor can grade the exams in 63 hours.

8. The pack will weigh 13.6 pounds on the moon.

9. $P = \dfrac{k}{w}$

$125 = \dfrac{k}{4000}$

$500{,}000 = k$

$P = \dfrac{500{,}000}{6000} = 83\dfrac{1}{3}$

Light with a wavelength of 6000 angstroms

will produce a magnification power of $83\dfrac{1}{3}$ times.

10. The area of the ellipse is 216π square centimeters.

11. 6.2%

12. 10.5%

13. White

14. Dark blue and silver are tied for the least popular color.

15. 21.7%

16. Approximately 195,059 cars

Chapter 6 Practice Test

1. $\{x \mid x \text{ is a real number and } x \neq 3\}$

2. $\{x \mid x \text{ is a real number and } x \neq -5\}$

3. $-\dfrac{2}{5}$

4. $\dfrac{x+3}{2x^2 - 3x + 5}$

5. $\dfrac{y-3}{8}$

6. $\dfrac{p^3 r^3}{2q}$

7. $\dfrac{6x^2 - 32x + 20}{(x+2)(x-7)}$

8. $\dfrac{4x-5}{(x+3)(x+1)}$

9. $\dfrac{21}{4}$

10. $6ab + \dfrac{3}{c} - \dfrac{2c}{a}$

11. $x^2 - 9$

12. $3x^3 - 9x^2 + 20x - 20 + \dfrac{29}{x+2}$

13. $(f + g)(x) = 10 - 3x$; $(g \cdot f)(x) = 40x - 28x^2$

14. $\left\{ -\dfrac{23}{56} \right\}$

15. $\left\{ \dfrac{2}{3} \right\}$

16. $\{-14\}$

17. The number is 7.

18. It will take $5\dfrac{5}{6}$ hours together.

19. $w = 4.8$

20. $P = 6120$

Chapter 7

Section 7.1
1. 40
2. −25
3. $7x^3y^2$
4. $3ab^3$
5. p^3qr^5
6. $-t$
7. $q - 4$
8. Not a real number
9. $\{x \mid x \text{ is a real number and } x \geq 0\}$

x	$f(x)$
0	−4
1	−3
4	−2
9	−1

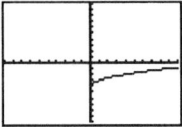

10. $\{x \mid x \text{ is a real number and } x \geq -3\}$

x	$f(x)$
−3	0
0	$\sqrt{3}$
1	2
6	3

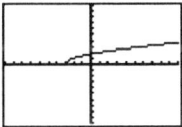

11. $\{x \mid x \text{ is a real number}\}$

x	$f(x)$
−6	−2
−1	$-\sqrt[3]{3}$
3	1
10	2

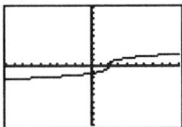

12. $\{x \mid x \text{ is a real number}\}$

x	$f(x)$
−1	2
0	3
1	4
8	5

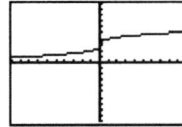

13. $\{x \mid x \text{ is a real number and } x \geq 7\}$

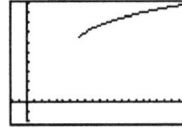

14. $\{x \mid x \text{ is a real number}\}$

Section 7.2
1. $(9m)^{3/2} = \left(\sqrt{9m}\right)^3 = \left(3\sqrt{m}\right)^3 = 27m\sqrt{m}$
2. −2
3. $5\sqrt[5]{a^4}$
4. $\sqrt[3]{(x+7)^2}$
5. $\dfrac{1}{x^{3/4}}$
6. 9
7. $\dfrac{2m^{5/2}}{5}$
8. $\dfrac{3}{q^{1/7}}$
9. $q^{2/7}$
10. $4^{1/5}r^{3/5}$
11. $\sqrt{(3r-10)^4} = (3r-10)^{4/2}$
$$= (3r-10)^2$$
$$= 9r^2 - 60r + 100$$
12. $4a^{1/2} - 5b^{1/3}$
13. $q^{3/5}q^{7/5} = q^{3/5+7/5} = q^{10/5} = q^2$
14. $\dfrac{b^{3/4}}{a^{1/3}}$
15. $xy^3z^{3/2}$
16. $pq^{3/4}r^4$
17. $x^{2/5}(2x^{2/5} - 3)$
18. $5x^{4/3}(x+2)$
19. 5.8088
20. 55.9017

Section 7.3

1. $\sqrt[3]{9} \cdot \sqrt[3]{3} = \sqrt[3]{9 \cdot 3} = \sqrt[3]{27} = 3$

2. $y\sqrt[4]{x^3}$

3. $\sqrt[5]{10a^3b^4}$

4. $\sqrt[3]{9}$

5. $\dfrac{\sqrt{60x^4}}{\sqrt{15x^3}} = \sqrt{\dfrac{60x^4}{15x^3}} = \sqrt{4x} = 2\sqrt{x}$

6. $q\sqrt[3]{p}$

7. $\sqrt[3]{81} = \sqrt[3]{27 \cdot 3} = \sqrt[3]{27} \cdot \sqrt[3]{3} = 3\sqrt[3]{3}$

8. $4\sqrt{10}$

9. $xz\sqrt[5]{40x^2y^4}$

10. $-5a^5b^3c^2\sqrt[3]{bc^2}$

11. $\dfrac{x\sqrt{x}}{2y}$

12. $\dfrac{a^2b^3\sqrt[3]{b^2}}{4}$

13. $\sqrt{3} \cdot \sqrt[3]{4} = 3^{1/2} \cdot 4^{1/3} = 3^{3/6} \cdot 4^{2/6} = \sqrt[6]{27} \cdot \sqrt[6]{16} = \sqrt[6]{432}$

14. $\sqrt[15]{62,208x^3}$

15. Area $= 5\pi\sqrt{5^2 + 12^2} = 5\pi\sqrt{169} = 65\pi$ square inches

16. 1,000,000 copies

Section 7.4

1. $\sqrt{50} - \sqrt{72} = \sqrt{2}\sqrt{25} - \sqrt{2}\sqrt{36}$
$$= 5\sqrt{2} - 6\sqrt{2} = -\sqrt{2}$$

2. $5\sqrt{5}$

3. $7a\sqrt[4]{ab} - \sqrt[4]{16a^5b} = 7a\sqrt[4]{ab} - 2a\sqrt[4]{ab} = 5a\sqrt[4]{ab}$

4. $13x\sqrt[3]{2xy^2}$

5. $\dfrac{6\sqrt{5}}{5} - \dfrac{\sqrt{5}}{5} = \dfrac{6\sqrt{5} - \sqrt{5}}{5} = \dfrac{5\sqrt{5}}{5} = \sqrt{5}$

6. $\dfrac{23\sqrt{3}}{12}$

7. $\dfrac{\sqrt[3]{32x^2}}{2} + \dfrac{\sqrt[3]{4x^2}}{6} = \dfrac{2\sqrt[3]{4x^2}}{2} + \dfrac{\sqrt[3]{4x^2}}{6}$
$$= \dfrac{6\sqrt[3]{4x^2}}{6} + \dfrac{\sqrt[3]{4x^2}}{6}$$
$$= \dfrac{7\sqrt[3]{4x^2}}{6}$$

8. $\dfrac{x\sqrt{x}(2-x)}{10}$

9. $\sqrt{3}\left(\sqrt{6} + \sqrt{10}\right) = \sqrt{3}\sqrt{6} + \sqrt{3}\sqrt{10}$
$$= \sqrt{18} + \sqrt{30}$$
$$= 3\sqrt{2} + \sqrt{30}$$

10. $13 - 4\sqrt{10}$

11. $\sqrt{10x}\left(\sqrt{2} + \sqrt{x}\right) = \sqrt{10x}\sqrt{2} + \sqrt{10x}\sqrt{x}$
$$= \sqrt{20x} + \sqrt{10x^2}$$
$$= 2\sqrt{5x} + x\sqrt{10}$$

12. $4x - 3$

13. $\left(2\sqrt{2y} + \sqrt{3y}\right)\left(2 - \sqrt{y}\right)$
$$= 2\left(2\sqrt{2y}\right) - 2\sqrt{2y}\sqrt{y} + 2\sqrt{3y} - \sqrt{3y}\sqrt{y}$$
$$= 4\sqrt{2y} - 2\sqrt{2y^2} + 2\sqrt{3y} - \sqrt{3y^2}$$
$$= 4\sqrt{2y} - 2y\sqrt{2} + 2\sqrt{3y} - y\sqrt{3}$$

14. $a^2\sqrt{2} - 3\sqrt{2}$

15. $\sqrt[3]{3x}\left(\sqrt[3]{9x^2} + \sqrt[3]{2}\right) = \sqrt[3]{3x}\sqrt[3]{9x^2} + \sqrt[3]{3x}\sqrt[3]{2}$
$$= \sqrt[3]{27x^3} + \sqrt[3]{6x} = 3x + \sqrt[3]{6x}$$

16. $x + \sqrt[3]{x^2} - \sqrt[3]{x} - 1$

17. $2\sqrt{5} + \sqrt{45} + \sqrt{125} = 2\sqrt{5} + 3\sqrt{5} + 5\sqrt{5}$
$$= 10\sqrt{5} \text{ centimeters}$$

18. $4\sqrt[3]{36}$ square inches

Section 7.5

1. $\dfrac{\sqrt[3]{4}}{\sqrt[3]{9}} = \dfrac{\sqrt[3]{4}}{\sqrt[3]{3^2}} \cdot \dfrac{\sqrt[3]{3}}{\sqrt[3]{3}} = \dfrac{\sqrt[3]{12}}{\sqrt[3]{27}} = \dfrac{\sqrt[3]{12}}{3}$

2. $\dfrac{\sqrt{35b}}{5b}$

3. $\dfrac{12}{1-\sqrt{3}} = \dfrac{12}{1-\sqrt{3}} \cdot \dfrac{1+\sqrt{3}}{1+\sqrt{3}}$
$$= \dfrac{12 + 12\sqrt{3}}{1^2 - \left(\sqrt{3}\right)^2}$$
$$= \dfrac{12 + 12\sqrt{3}}{1 - 3}$$
$$= \dfrac{12 + 12\sqrt{3}}{-2}$$
$$= -6 - 6\sqrt{3}$$

4. $-24 + 6\sqrt{22}$

5.
$$\frac{6y}{\sqrt{y}+4} = \frac{6y}{\sqrt{y}+4} \cdot \frac{\sqrt{y}-4}{\sqrt{y}-4}$$
$$= \frac{6y\sqrt{y}-24y}{y-16}$$

6. $\dfrac{3-2y\sqrt{3}+y^2}{3-y^2}$

7.
$$\frac{\sqrt{2z}+3}{\sqrt{2z}-15} = \frac{\sqrt{2z}+3}{\sqrt{2z}-15} \cdot \frac{\sqrt{2z}+15}{\sqrt{2z}+15}$$
$$= \frac{2z+18\sqrt{2z}+45}{2z-225}$$

8. $-\dfrac{3+\sqrt{3}}{2}$

9. $\sqrt{\dfrac{20}{3}} = \dfrac{\sqrt{20}}{\sqrt{3}} \cdot \dfrac{\sqrt{20}}{\sqrt{20}} = \dfrac{20}{\sqrt{60}}$

10. $\dfrac{2}{\sqrt[3]{18}}$

11. $\dfrac{\sqrt{5a}}{15} = \dfrac{\sqrt{5a}}{15} \cdot \dfrac{\sqrt{5a}}{\sqrt{5a}} = \dfrac{5a}{15\sqrt{5a}} = \dfrac{a}{3\sqrt{5a}}$

12. $\dfrac{6z}{5\sqrt[3]{12z}}$

13. $\dfrac{\sqrt{3x}}{x-\sqrt{3x}} = \dfrac{\sqrt{3x}}{x-\sqrt{3x}} \cdot \dfrac{\sqrt{3x}}{\sqrt{3x}} = \dfrac{3x}{x\sqrt{3x}-3x}$

14. $\dfrac{4x^2}{5x\sqrt{2x}+2x}$

15. $\dfrac{\sqrt{3}-y}{\sqrt{3}+y} = \dfrac{\sqrt{3}-y}{\sqrt{3}+y} \cdot \dfrac{\sqrt{3}+y}{\sqrt{3}+y} = \dfrac{3-y^2}{3+2y\sqrt{3}+y^2}$

16. $\dfrac{2z-9}{2z-18\sqrt{2z}+45}$

17.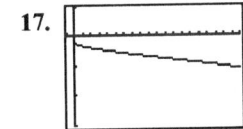

18.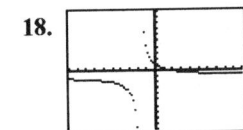

Section 7.6

1.
$$\sqrt{3x-8} = 8$$
$$3x-8 = 64$$
$$3x = 72$$
$$x = 24$$
$$\{24\}$$

2. $\{13\}$

3.
$$t+\sqrt{t+2} = 0$$
$$t = -\sqrt{t+2}$$
$$t^2 = t+2$$
$$t^2-t-2 = 0$$
$$(t-2)(t+1) = 0$$
$$t = -1, 2 \text{ (extraneous)}$$
$$\{-1\}$$

4. $\{3\}$

5.
$$\sqrt[3]{2x+6}-1 = 7$$
$$\sqrt[3]{2x+6} = 8$$
$$2x+6 = 512$$
$$2x = 506$$
$$x = 253$$
$$\{253\}$$

6. $\{6\}$

7.
$$2z+3 = \sqrt{5z+9}$$
$$(2z+3)^2 = 5z+9$$
$$4z^2+12z+9 = 5z+9$$
$$4z^2+7z = 0$$
$$z(4z+7) = 0$$
$$z = 0, -\frac{7}{4} \text{ extraneous}$$
$$\{0\}$$

8. $\{\ \}$

9.
$$\sqrt{x-4}-\sqrt{x+4} = -2$$
$$\sqrt{x-4} = \sqrt{x+4}-2$$
$$x-4 = x+4-4\sqrt{x+4}+4$$
$$4\sqrt{x+4} = 12$$
$$16(x+4) = 144$$
$$x+4 = 9$$
$$x = 5$$
$$\{5\}$$

10. $\{2, 3\}$

11.
$$10^2+5^2 = c^2$$
$$100+25 = c^2$$
$$125 = c^2$$
$$c = \sqrt{125} = 5\sqrt{5} \text{ meters}$$

12. $\sqrt{1045}$ inches

13.
$$8 = 5\sqrt{6 - 0.8x}$$
$$64 = 25(6 - 0.8x)$$
$$64 = 150 - 20x$$
$$-86 = -20x$$
$$x = 4.3$$

The greatest price that can be charged so that at least 800,000 magazines will be sold annually is $4.30.

14. Approximately 8.9 feet

Section 7.7

1. $(1+i) - (3-i) = 1 + i - 3 + i = -2 + 2i$

2. $-7 - 4i$

3. $(18 + 7i) + (-15 + 2i) = 18 + 7i - 15 + 2i = 3 + 9i$

4. $-3 + 4i$

5. $(-3i)(20i) = -60i^2 = -60(-1) = 60$

6. -132

7. $2i(7 + 15i) = 14i + 30i^2 = 14i - 30 = -30 + 14i$

8. $76 + 32i$

9. $(2 - 7i)(2 + 7i) = 4^2 - (7i)^2 = 16 - 49i^2$
$$= 16 + 49 = 65$$

10. $1 - 3i$

11. $\dfrac{1 + 9i}{3 - 2i} = \dfrac{1 + 9i}{3 - 2i} \cdot \dfrac{3 + 2i}{3 + 2i}$
$$= \dfrac{3 + 29i + 18i^2}{9 - 4i^2}$$
$$= \dfrac{3 + 29i - 18}{9 + 4}$$
$$= \dfrac{-15 + 29i}{13}$$
$$= -\dfrac{15}{13} + \dfrac{29}{13}i$$

12. $\dfrac{17}{26} + \dfrac{59}{26}i$

13. $\sqrt{-6} \cdot \sqrt{-8} = i\sqrt{6} \cdot i\sqrt{8} = i^2\sqrt{48} = -4\sqrt{3}$

14. $-2\sqrt{30}$

15. $\sqrt{-7} \cdot \sqrt{8} = i\sqrt{7} \cdot \sqrt{8} = i\sqrt{56} = 2i\sqrt{14}$

16. $\sqrt{15}$

17. $i^{13} = i^{12} \cdot i = \left(i^4\right)^3 i = 1i = i$

18. -1

19. $i^{28} = \left(i^4\right)^7 = 1^7 = 1$

20. i

Chapter 7 Practice Test

1. $7\sqrt{7}$

2. $-x^{11}$

3. $\dfrac{1}{4}$

4. 4

5. $\dfrac{27x^3}{8}$

6. $-q^8 r^{13}$

7. $\dfrac{6y^{1/12}z^{3/8}}{x^{1/6}}$

8. $y^{25/6} - y$

9. $13ab$

10. $|7x|$

11. $\dfrac{5\sqrt{3x}}{3x}$

12. $\dfrac{30 + 17\sqrt{y} + 2y}{36 - y}$

13. $\dfrac{\sqrt[3]{5xyz^2}}{z}$

14. $-7y\sqrt[3]{3y^2}$

15. $10 - \sqrt{26}$

16. $2a - 8\sqrt{2a} + 16$

17. $\sqrt{21} - \sqrt{3} + 5\sqrt{7} - 5$

18. 2

19. 19.026

20. 0.014

21. $\{3\}$

22. $\{\ \ \}$

23. $\{5\}$

24. $3i$

25. $-3i\sqrt{2}$

26. $5 + 3i$

27. 18

28. $-24 - 70i$

29. $\dfrac{7}{4} + \dfrac{9}{4}i$

30. $x = 6\sqrt{2}$

31. $\{x \mid x \text{ is a real number and } x \geq 2\}$

x	$f(x)$
2	0
2.5	1
4	2
6	$2\sqrt{2}$

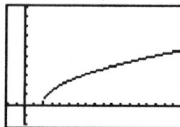

The solution set is $\left\{\dfrac{-1+i\sqrt{3}}{2}, \dfrac{-1-i\sqrt{3}}{2}\right\}$.

8. $\left\{\dfrac{-7+\sqrt{17}}{2}, \dfrac{-7-\sqrt{17}}{2}\right\}$

9. $2a^2 - 6a - 20 = 0$

$a = 2, b = -6, c = -20$

$x = \dfrac{-(-6)\pm\sqrt{(-6)^2 - 4(2)(-20)}}{2(2)}$

$= \dfrac{6\pm\sqrt{196}}{4}$

$= \dfrac{6\pm 14}{4}$

The solution set is $\{5, -2\}$.

10. $\left\{-1+i\sqrt{19}, -1-i\sqrt{19}\right\}$

11. $3t^2 - 9t + 36 = 0$

$a = 3, b = -9, c = 36$

$x = \dfrac{-(-9)\pm\sqrt{(-9)^2 - 4(3)(36)}}{2(3)}$

$= \dfrac{9\pm\sqrt{-351}}{6}$

$= \dfrac{9\pm 3i\sqrt{39}}{6}$

The solution set is $\left\{\dfrac{3+i\sqrt{39}}{2}, \dfrac{3-i\sqrt{39}}{2}\right\}$.

12. $\left\{2+\sqrt{19}, 2-\sqrt{19}\right\}$

13. $2x^2 - 9x - 17 = 0$

$b^2 - 4ac = (-9)^2 - 4(2)(-17) = 217$

Because the discriminant is positive, the equation has two real solutions.

14. Two real solutions

15. $3y^2 + y + 4 = 0$

$b^2 - 4ac = 1^2 - 4(3)(4) = -47$

Because the discriminant is negative, the equations has two complex solutions.

16. One real solution

17. $0 = -16t^2 - 15t + 546$

$a = -16, b = -15, c = 546$

$x = \dfrac{-(-15)\pm\sqrt{(-15)^2 - 4(-16)(546)}}{2(-16)}$

$x \approx -6.33, 5.39$

The ball will hit the ground after

approximately 5.39 seconds.

18. (a) 1987

(b) $6.9 billion

(c) $7.339 billion; this is somewhat close to the graph value of nearly $7.5 billion

(d) In 1986 and 1988; these are very close to the graph values

Section 8.3

1. $\dfrac{1}{x} - \dfrac{4}{x-1} = 2$

$1(x-1) - 4x = 2x(x-1)$

$x - 1 - 4x = 2x^2 - 2x$

$0 = 2x^2 + x + 1$

$x = \dfrac{-1\pm\sqrt{1 - 4(2)(1)}}{4}$

$= \dfrac{-1\pm i\sqrt{7}}{4}$

The solution set is $\left\{\dfrac{-1+i\sqrt{7}}{4}, \dfrac{-1-i\sqrt{7}}{4}\right\}$.

2. $\left\{-4+\sqrt{19}, -4-\sqrt{19}\right\}$

3. $y^4 + 6y^2 + 5 = 0$

$(y^2 + 5)(y^2 + 1) = 0$

$y^2 + 5 = 0$

$y^2 = -5$

$y = \pm i\sqrt{5}$

$y^2 + 1 = 0$

$y^2 = -1$

$y = \pm i$

The solution set is $\left\{i, -i, i\sqrt{5}, -i\sqrt{5}\right\}$.

4. $\left\{i\sqrt{3}, -i\sqrt{3}, 2i, -2i\right\}$

5. $(2x-4)^2 - 3(2x-4) + 2 = 0$

$y^2 - 3y + 2 = 0$

$(y-2)(y-1) = 0$

$y = 1, 2$

$2x - 4 = 1$	$2x - 4 = 2$
$2x = 5$	$2x = 6$
$x = \dfrac{5}{2}$	$x = 3$

The solution set is $\left\{\dfrac{5}{2}, 3\right\}$.

6. $\left\{ \dfrac{-1+i\sqrt{7}}{2}, \dfrac{-1-i\sqrt{7}}{2} \right\}$

7. $x^{2/3} + x^{1/3} - 6 = 0$

$$m^2 + m - 6 = 0$$

$$(m+3)(m-2) = 0$$

$$m = -3, 2$$

$$x^{1/3} = -3 \qquad x^{1/3} = 2$$

$$x = (-3)^3 \qquad x = 2^3$$

$$x = -27 \qquad x = 8$$

The solution set is $\{8, -27\}$.

8. $\{-8, 64\}$

9. $\dfrac{2}{2x-3} + \dfrac{1}{x} = \dfrac{1}{5}$

$$10x + 5(2x - 3) = x(2x - 3)$$

$$10x + 10x - 15 = 2x^2 - 3x$$

$$0 = 2x^2 - 23x + 15$$

$$x = \dfrac{23 \pm \sqrt{409}}{4}$$

Printing press #2 can complete the run in 10.81 hours and printing press #1 can complete the run in $10.81 - 1.5 = 9.31$ hours.

10. 91 days

11. No real solutions; the algebraic solution to Exercise 4 yielded four complex solutions. These will not appear as x-intercepts on a graph because only real solutions appear as x-intercepts on a graph.

12. $\{2.50, 3.00\}$; this is the same solution as was obtained algebraically.

Section 8.4

1. $(2x - 3)(x + 1) > 0$

$$(2x - 3)(x + 1) = 0$$

$$x = \dfrac{3}{2}, -1$$

The solution set is $(-\infty, -1) \cup \left(\dfrac{3}{2}, \infty \right)$.

2. $(-\infty, 7) \cup (9, \infty)$

3. $(x + 9)(3x - 5) < 0$

$$(x + 9)(3x - 5) = 0$$

$$x = -9, \dfrac{5}{3}$$

The solution set is $\left(-9, \dfrac{5}{3} \right)$.

4. $[-5, 3]$

5. $x^2 + 3x - 10 \le 0$

$$x^2 + 3x - 10 = 0$$

$$(x + 5)(x - 2) = 0$$

$$x = -5, 2$$

The solution set is $[-5, 2]$.

6. $(-\infty, -3] \cup [5, \infty)$

7. $(x + 8)(x + 1)(x - 3) \ge 0$

$$(x + 8)(x + 1)(x - 3) = 0$$

$$x = -8, -1, 3$$

The solution set is $[-8, -1] \cup [3, \infty)$.

8. $(-\infty, -4] \cup [1, 5]$

9. $(t^2 - 3)(t + 11) < 0$

$$(t^2 - 3)(t + 11) = 0$$

$$t = \pm\sqrt{3}, -11$$

The solution set is $(-\infty, -11) \cup \left(-\sqrt{3}, \sqrt{3} \right)$.

10. $(-\infty, -9) \cup (-\sqrt{10}, 0) \cup (\sqrt{10}, \infty)$

11. $\dfrac{x - 3}{x + 10} > 0$

$$x + 10 = 0$$

$$x = -10$$

$$x - 3 = 0$$

$$x = 3$$

The solution set is $(-\infty, -10) \cup (3, \infty)$.

12. $[-2, 7)$

13. $\dfrac{2}{x - 2} \ge 3$

$$x - 2 = 0$$

$$x = 2$$

$$2 = 3(x - 2)$$

$$2 = 3x - 6$$

$$0 = 3x - 8$$

$$x = \dfrac{8}{3}$$

The solution set is $\left(2, \dfrac{8}{3} \right]$.

14. $(-\infty, -4) \cup \left(-\dfrac{19}{5}, \infty \right)$

15.

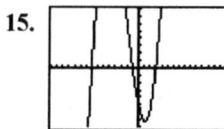

The graph is greater than 0 when x is between -8 and -1, and between 3 and ∞.

16.

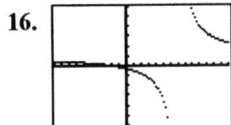

The graph is less than 0 when x is in the interval $[-2, 7)$.

Section 8.5
1. $(0, 0)$
2. $(0, -4)$
3. $(-3, 0)$
4. $(2, 3)$
5.

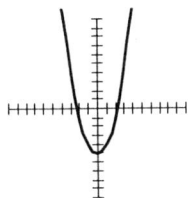

Vertex: $(0, -5)$; axis of symmetry: $x = 0$

6.

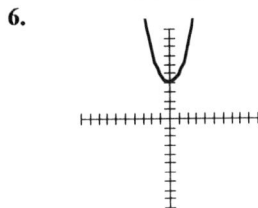

Vertex: $(0, 4)$; axis of symmetry: $x = 0$

7.

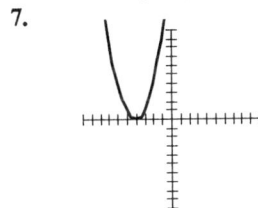

Vertex: $(-4, 0)$; axis of symmetry: $x = -4$

8.

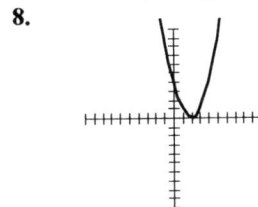

Vertex: $(2, 0)$; axis of symmetry: $x = 2$

9.

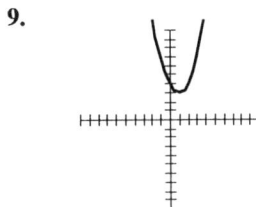

Vertex: $(1, 3)$; axis of symmetry: $x = 1$

10.

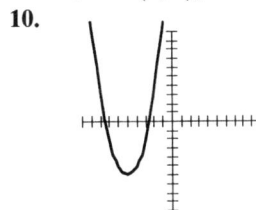

Vertex: $(-5, -6)$; axis of symmetry: $x = -5$

11.

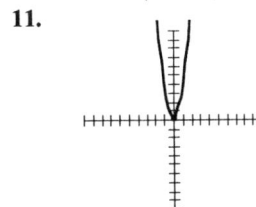

Vertex: $(0, 0)$; axis of symmetry: $x = 0$

12.

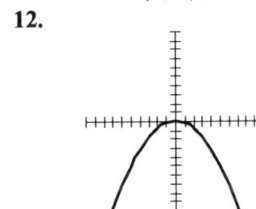

Vertex: $(0, 0)$; axis of symmetry: $x = 0$

13.

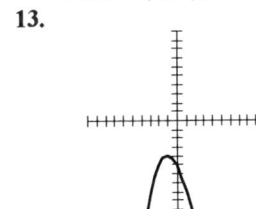

Vertex: $(-1, -4)$; axis of symmetry: $x = -1$

14.

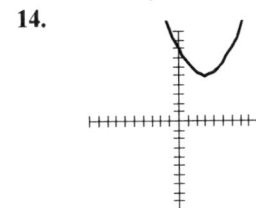

Vertex: $(3, 5)$; axis of symmetry: $x = 3$

15.
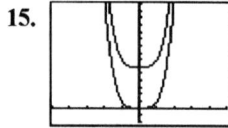

16.
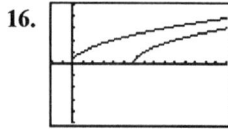

17.
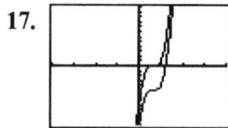

18.
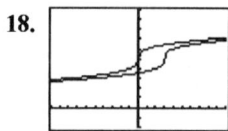

Section 8.6

1.
$$f(x) = x^2 + 8x + 10$$
$$y = x^2 + 8x + 10$$
$$y - 10 = x^2 + 8x$$
$$y - 10 + 16 = x^2 + 8x + 16$$
$$y + 6 = (x + 4)^2$$
$$y = (x + 4)^2 - 6$$
Vertex: (–4, –6)

2. $y = -(x - 2)^2 + 9$
Vertex: (2, 9)

3.
$$h(x) = -2x^2 + 12x + 11$$
$$y = -2x^2 + 12x + 11$$
$$y - 11 = -2(x^2 - 6x)$$
$$y - 11 - 18 = -2(x^2 - 6x + 9)$$
$$y - 29 = -2(x - 3)^2$$
$$y = -2(x - 3)^2 + 29$$
Vertex: (3, 29)

4.
$$y = 3\left(x + \frac{5}{2}\right)^2 - \frac{67}{4}$$
Vertex: $\left(-\frac{5}{2}, -\frac{67}{4}\right)$

5. $f(x) = 3x^2 + 36x - 1$
$$\frac{-b}{2a} = \frac{-36}{2(3)} = -6$$
$$f(-6) = 3(-6)^2 + 36(-6) - 1 = -109$$

Vertex: (–6, –109)

6. Vertex: $\left(\frac{5}{2}, -\frac{19}{2}\right)$

7. $h(x) = x^2 - 18x - 4$
$$\frac{-b}{2a} = \frac{-(-18)}{2(1)} = 9$$
$$h(9) = 9^2 - 18(9) - 4 = -85$$
Vertex: (9, –85)

8. Vertex: (–14, –377)

9.
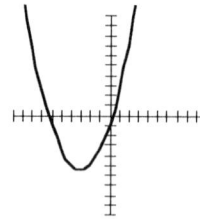

10.
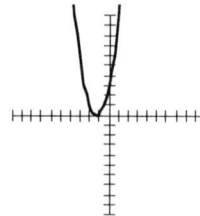

11.
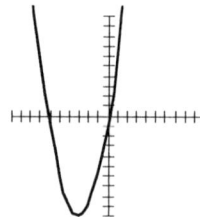

12.
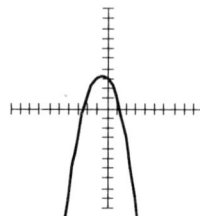

13.
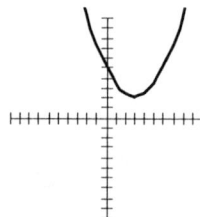

14.

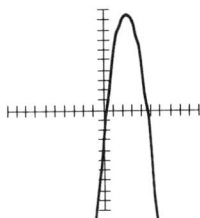

15. $h(t) = -16t^2 + 40t + 50$

$$\frac{-b}{2a} = \frac{-40}{2(-16)} = \frac{40}{32} = \frac{5}{4}$$

$$h\left(\frac{5}{4}\right) = -16\left(\frac{5}{4}\right)^2 + 40\left(\frac{5}{4}\right) + 50 = 75$$

The maximum height of the projectile is 75 feet.

16. (a) 600 compact discs

(b) $20,000

17. $f(x) = x(100 - x) = 100x - x^2$

$$\frac{-b}{2a} = \frac{-100}{2(-1)} = 50$$

$100 - 50 = 50$

The numbers are 50 and 50.

18.

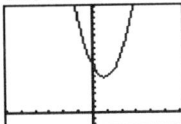

Vertex: (0.81, 5.88)

19.

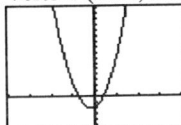

Vertex: (−0.18, −2.00)

Chapter 8 Practice Test

1. $\left\{-1, \frac{11}{3}\right\}$

2. $\left\{3 + 3\sqrt{2}, 3 - 3\sqrt{2}\right\}$

3. $\left\{-4 + \sqrt{30}, -4 - \sqrt{30}\right\}$

4. $\left\{\frac{-9 + 5\sqrt{3}}{2}, \frac{-9 - 5\sqrt{3}}{2}\right\}$

5. $\left\{-\frac{2}{5}, -1\right\}$

6. $\left\{-1 + \sqrt{10}, -1 - \sqrt{10}\right\}$

7. $\left\{\frac{-7 + i\sqrt{11}}{6}, \frac{-7 - i\sqrt{11}}{6}\right\}$

8. $\left\{i\sqrt{5}, -i\sqrt{5}, \sqrt{3}, -\sqrt{3}\right\}$

9. $\left\{2, -2, 2i, -2i, i\sqrt{2}, -i\sqrt{2}\right\}$

10. $\{3, 11\}$

11. $\left\{1 + i\sqrt{11}, 1 - i\sqrt{11}\right\}$

12. $\left\{2 + \frac{\sqrt{46}}{2}, 2 - \frac{\sqrt{46}}{2}\right\}$

13. $\left(\frac{4}{3}, -2\right)$

14. $(-\infty, -9) \cup (-3, 0) \cup (9, \infty)$

15. $(-\infty, -5) \cup \left[-\frac{1}{4}, 5\right)$

16.

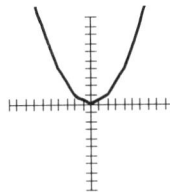

Vertex: (0, 0)

17.

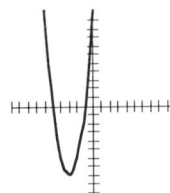

Vertex: (−3, −8)

18.

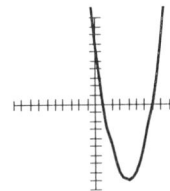

Vertex: (4, −9)

19.

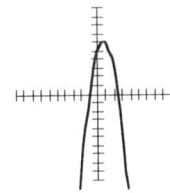

Vertex: $\left(\frac{3}{4}, \frac{49}{8}\right)$

20. $3 + 3\sqrt{17}$ or 15.4 feet

21. 13.7 hours and 10.7 hours

Chapter 9

Section 9.1

1.

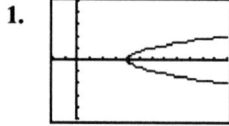

Vertex: (4, 0)

2.

Vertex: (–3, 1)

3.

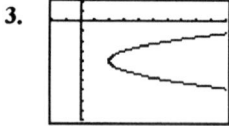

Vertex: (2, –4)

4.

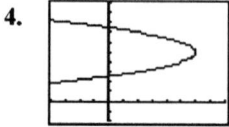

Vertex: (6, 5)

5. (3, 5), (6, 1)

$$d = \sqrt{(3-6)^2 + (5-1)^2}$$
$$= \sqrt{(-3)^2 + (4)^2}$$
$$= \sqrt{9+16}$$
$$= \sqrt{25}$$
$$= 5$$

$$\text{midpoint} = \left(\frac{3+6}{2}, \frac{5+1}{2}\right)$$
$$= \left(\frac{9}{2}, \frac{6}{2}\right)$$
$$= \left(\frac{9}{2}, 3\right)$$

6. $\sqrt{29}$; $\left(1, \frac{3}{2}\right)$

7. (–7, 2), (3, –8)

$$d = \sqrt{(-7-3)^2 + (2-(-8))^2}$$
$$= \sqrt{(-10)^2 + (10)^2}$$
$$= \sqrt{100+100}$$
$$= \sqrt{200}$$
$$= 10\sqrt{2}$$

$$\text{midpoint} = \left(\frac{-7+3}{2}, \frac{2+(-8)}{2}\right)$$
$$= \left(\frac{-4}{2}, \frac{-6}{2}\right)$$
$$= (-2, -3)$$

8. 13, $\left(-\frac{15}{2}, 3\right)$

9. (0, 0); 1

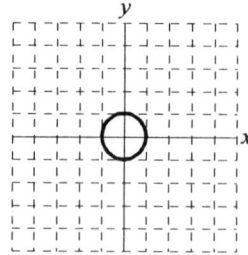

10. (0, –2); 3

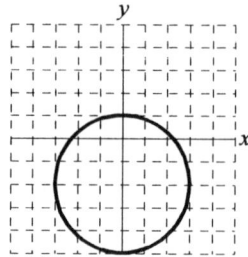

11. (1, –1); 2

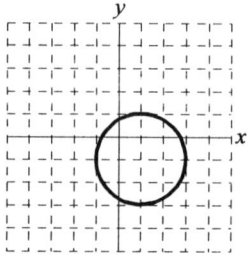

12. (–2, –1); 1

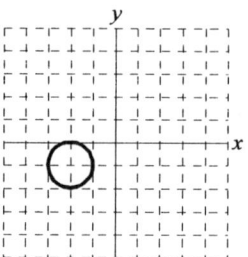

13. (–1, 0); 1

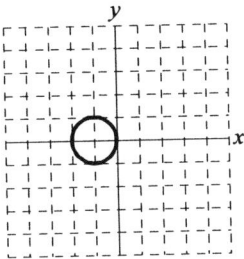

14. (0, −2); 2

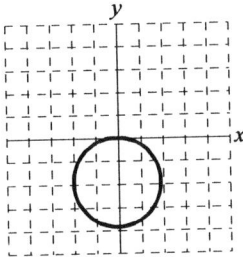

15. Center: (5, −3); radius: 5
$(x - 5)^2 + (y + 3)^2 = 25$

16. Center: (−1, 10); radius: 7
$(x + 1)^2 + (y - 10)^2 = 49$

17. (0, 0), (3, 4), and (5, −2)
$d = \sqrt{(3 - 0)^2 + (4 - 0)^2} = 5$
$d = \sqrt{(5 - 0)^2 + (-2 - 0)^2} = \sqrt{29}$
$d = \sqrt{(5 - 3)^2 + (-2 - 4)^2} = 2\sqrt{10}$

No, the triangle is not isosceles.

18. Yes, the triangle is isosceles.

19.

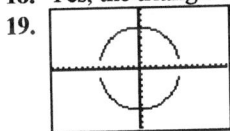

20.

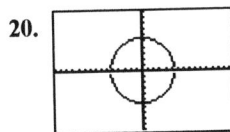

Section 9.2

1.

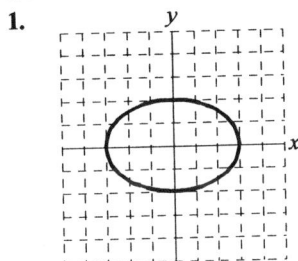

2.

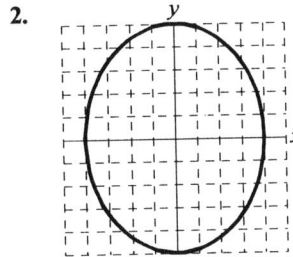

3.

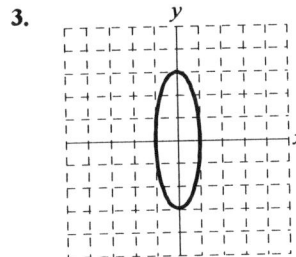

4.

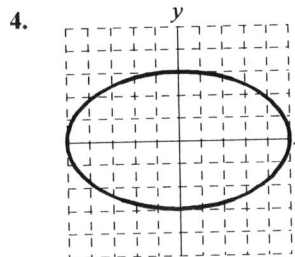

5.

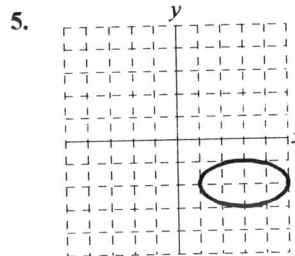

6.

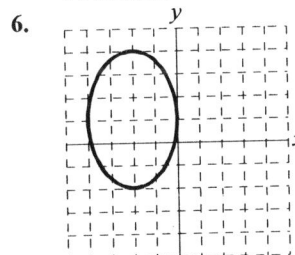

7.

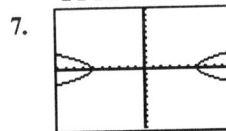

8.

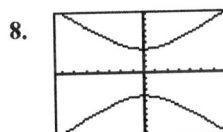

9.

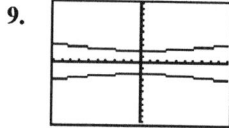

10.

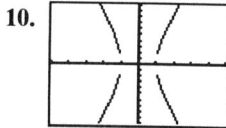

11. Parabola
12. Ellipse
13. Circle
14. Ellipse
15. Circle
16. Hyperbola
17.

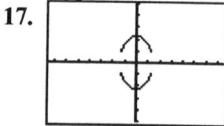

18.

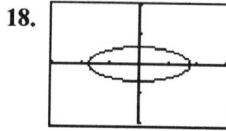

Section 9.3

1. $\begin{cases} y = 2x \\ y = x^2 - x \end{cases}$

$2x = x^2 - x$

$0 = x^2 - 3x$

$0 = x(x - 3)$

$x = 0, 3$

$y = 2(0) = 0$

$y = 2(3) = 6$

The solution set is $\{(0, 0), (3, 6)\}$.

2. $\{(-3, 1), (-4, 0)\}$

3. $\begin{cases} y = 3 \\ 4y^2 - x^2 = 20 \end{cases}$

$4(3)^2 - x^2 = 20$

$4(9) - x^2 = 20$

$36 - 20 = x^2$

$16 = x^2$

$x = \pm 4$

The solution set is $\{(4, 3), (-4, 3)\}$.

4. $\{(1, 0), (3, 4)\}$

5. $\begin{cases} x^2 + y^2 = 4 \\ y - x^2 = -2 \end{cases} \Rightarrow \begin{cases} x^2 + y^2 = 4 \\ \underline{-x^2 + y = -2} \end{cases}$

$y^2 + y = 2$

$y^2 + y - 2 = 0$

$(y + 2)(y - 1) = 0$

$y = -2, 1$

$-2 - x^2 = -2$

$x^2 = 0 \Rightarrow x = 0$

$1 - x^2 = -2$

$x^2 = 3 \Rightarrow x = \pm\sqrt{3}$

The solution set is $\{(0, -2), (\sqrt{3}, 1),$ $(-\sqrt{3}, 1)\}$.

6. $\{(\sqrt{3}, 0), (-\sqrt{3}, 0), (\sqrt{5}, 2), (-\sqrt{5}, 2)\}$

7. $\begin{cases} x^2 + y^2 = 9 \\ x^2 + 9y^2 = 9 \end{cases} \Rightarrow \begin{cases} -x^2 - y^2 = -9 \\ \underline{x^2 + 9y^2 = 9} \end{cases}$

$8y^2 = 0$

$y = 0$

$x^2 + 0 = 9 \Rightarrow x = \pm 3$

The solution set is $\{(3, 0), (-3, 0)\}$.

8. $\left\{ \left(\dfrac{\sqrt{3}}{3}, \dfrac{\sqrt{2}}{3} \right), \left(\dfrac{\sqrt{3}}{3}, -\dfrac{\sqrt{2}}{3} \right), \left(-\dfrac{\sqrt{3}}{3}, \dfrac{\sqrt{2}}{3} \right), \right.$ $\left. \left(-\dfrac{\sqrt{3}}{3}, -\dfrac{\sqrt{2}}{3} \right) \right\}$

9. $\begin{cases} x^2 + y^2 = 125 \\ \underline{x^2 - y^2 = 117} \end{cases}$

$2x^2 = 242$

$x^2 = 121$

$x = \pm 11$

$(11)^2 + y^2 = 125$

$121 + y^2 = 125$

$y^2 = 4$

$y = \pm 2$

The numbers are 11 and 2, 11 and -2, -11 and 2, and -11 and -2.

10. The numbers are 3 and 8, and -3 and -8.

11. $\{(2,78, 1.5), (-2.78, 1.5), (-2.78, -1.5),$ $(2.78, -1.5)\}$

12. $\{(-1.87, 1.51), (0.43, 3.34)\}$

Section 9.4

1.

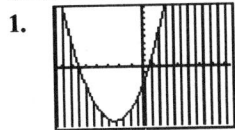

2.

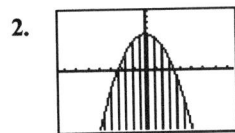

3.

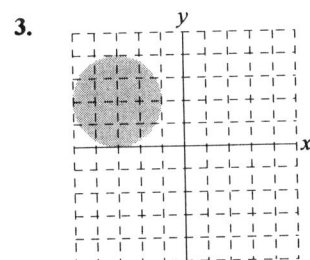

4.

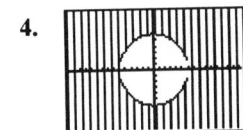

5.

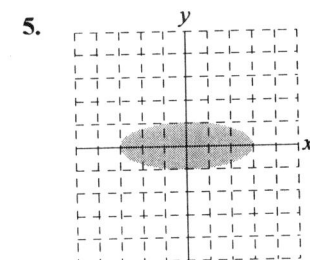

6.

7.

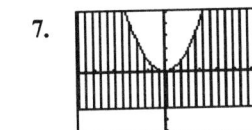

8.

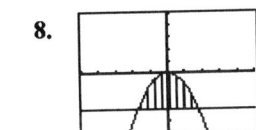

9.

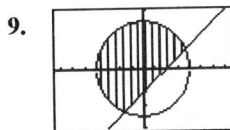

10.

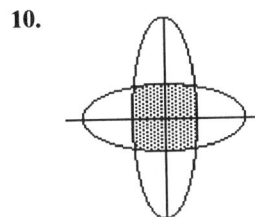

Chapter 9 Practice Test

1. 13

2. $2\sqrt{29}$

3. $\sqrt{53}$

4. $(5, -4)$

5. $\left(\dfrac{5}{8}, -\dfrac{2}{5}\right)$

6.

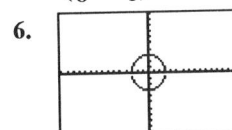

Center: $(0, 0)$, radius: 3, intercepts: $(3, 0)$, $(0, 3)$, $(-3, 0)$, $(0, -3)$

7.

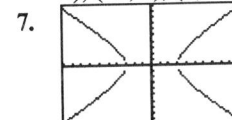

Center: $(0, 0)$, intercepts: $(3, 0)$, $(-3, 0)$, asymptotes: $y = x$ and $y = -x$

8.

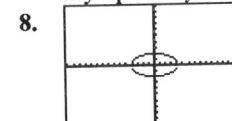

Center: $(0, 0)$, intercepts: $(4, 0)$, $(0, 2)$, $(-4, 0)$, $(0, -2)$

9.

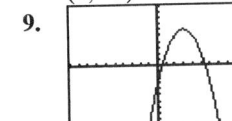

Vertex: $(3, 6)$, intercepts: $(0.55, 0)$, $(5.45, 0)$, $(0, -3)$

10.

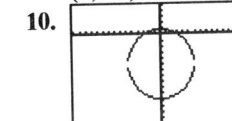

Center: $(0, -5)$, radius: $\sqrt{34}$, intercepts: $(-3, 0)$, $(3, 0)$, $(0, 0.831)$, $(0, -10.831)$

11.

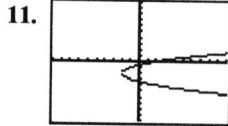

Vertex: $(-2, -2)$, intercepts: $(2, 0)$, $(0, -0.586)$, $(0, -3.414)$

12.

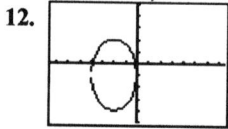

Center: $(-2, -1)$, intercepts: $(0, -1)$, $(-3.89, 0)$, $(-0.11, 0)$

13.

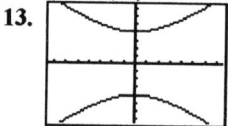

Center $(0, 0)$, intercepts: $(0, 4)$, $(0, -4)$, asymptotes: $y = x$ and $y = -x$

14. $\{(-8, 6), (8, -6)\}$

15. $\left\{ \left(9, \sqrt{17}\right), \left(9, -\sqrt{17}\right), \left(-9, \sqrt{17}\right), \left(-9, -\sqrt{17}\right) \right\}$

16. $\{(0, 4), (4, 4)\}$

17.

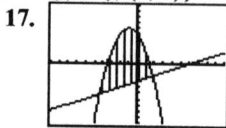

18.

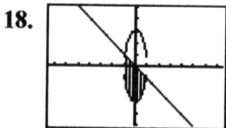

19.

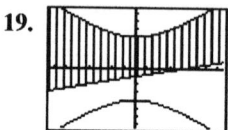

20. C

21. Height: 13 feet, width: 16 feet

Chapter 10

Section 10.1

1.
$$(f \circ g)(x) = f(g(x))$$
$$= f(5 + \sqrt{x})$$
$$= 4(5 + \sqrt{x})^2$$
$$= 4(25 + 10\sqrt{x} + x)$$
$$= 100 + 40\sqrt{x} + 4x$$

2. $(f \circ h)(x) = 400x^2 + 240x + 36$

3.
$$(g \circ h)(x) = g(h(x))$$
$$= g(10x + 3)$$
$$= 5 + \sqrt{10x + 3}$$

4. $(g \circ f)(x) = 5 + 2|x|$

5.
$$(h \circ f)(x) = h(f(x))$$
$$= h(4x^2)$$
$$= 10(4x^2) + 3$$
$$= 40x^2 + 3$$

6. $(h \circ g)(x) = 53 + 10\sqrt{x}$

7.
$$(g \circ h)(9) = 5 + \sqrt{10(9) + 3}$$
$$= 5 + \sqrt{93}$$

8. 11

9. No, this is not a function.

10. No, this function is not one-to-one.

11. Yes, this function is one-to-one.

12. No, this function is not one-to-one.

13.

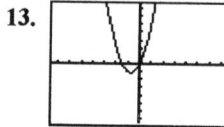

No, by the horizontal line test

14. Yes

15.

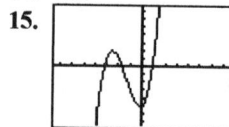

No, by the horizontal line test

16. Yes

17.
$$f(x) = 5 - 3x$$
$$x = 5 - 3y$$
$$3y = 5 - x$$
$$y = \frac{5 - x}{3}$$
$$f^{-1}(x) = \frac{5 - x}{3}$$

18. $f^{-1}(x) = 5x - 2$

19.
$$h(x) = \frac{x - 3}{x}$$
$$x = \frac{y - 3}{y}$$
$$xy = y - 3$$
$$xy - y = -3$$
$$y(x - 1) = -3$$
$$y = \frac{-3}{x - 1}$$

$$f^{-1}(x) = \frac{-3}{x-1}$$

20. $f^{-1}(x) = \frac{\sqrt{x+3}}{2}$

21. $f(x) = \frac{1}{2}x - 7$ and $f^{-1}(x) = 2x + 14$

$$(f \circ f^{-1})(x) = f(f^{-1}(x))$$
$$= f(2x + 14)$$
$$= \frac{1}{2}(2x + 14) - 7$$
$$= x + 7 - 7$$
$$= x$$

$$(f^{-1} \circ f)(x) = f^{-1}(f(x))$$
$$= f^{-1}\left(\frac{1}{2}x - 7\right)$$
$$= 2\left(\frac{1}{2}x - 7\right) + 14$$
$$= x - 14 + 14$$
$$= x$$

22. $(f \circ f^{-1})(x) = x$; $(f^{-1} \circ f)(x) = x$

23.
$$f(x) = \frac{1}{10}x - 7$$
$$x = \frac{1}{10}y - 7$$
$$x + 7 = \frac{1}{10}y$$
$$10x + 70 = y$$
$$f^{-1}(x) = 10x + 70$$

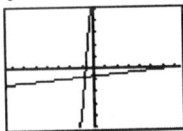

24. $f^{-1}(x) = \frac{x^3 + 7}{4}$

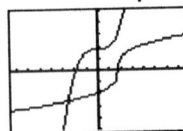

Section 10.2

1.

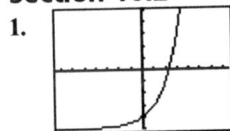

2.

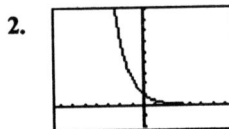

3.

4.

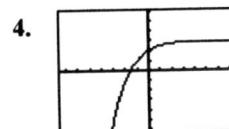

5. $32^x = 4$
$$2^{5x} = 2^2$$
$$5x = 2$$
$$x = \frac{2}{5}$$
The solution set is $\left\{\frac{2}{5}\right\}$.

6. $\left\{\frac{3}{2}\right\}$

7. $5^{2x+1} = 125$
$$5^{2x+1} = 5^3$$
$$2x + 1 = 3$$
$$2x = 2$$
$$x = 1$$
The solution set is $\{1\}$.

8. $\left\{\frac{10}{3}\right\}$

9. $125^{x+1} = 625^{2x-1}$
$$5^{3(x+1)} = 5^{4(2x-1)}$$
$$3x + 3 = 8x - 4$$
$$7 = 5x$$
$$x = \frac{7}{5}$$
The solution set is $\left\{\frac{7}{5}\right\}$.

10. $\{-2\}$

11. $27^{2x-8} = 9^{x-4}$
$$3^{3(2x-8)} = 3^{2(x-4)}$$
$$6x - 24 = 2x - 8$$
$$4x = 16$$
$$x = 4$$

The solution set is {4}.

12. $\left\{\dfrac{14}{5}\right\}$

13. $y = 3,900,000(2.7)^{0.037t}$

   $y = 3,900,000(2.7)^{0.037(5)}$

   $y = 3,900,000(2.7)^{0.185}$

   $\approx 4,686,697$

   There will be approximately 4,686,697 people living in Togo after 5 years.

14. Approximately 1,166,157 people

15.

   $A = P\left(1+\dfrac{r}{n}\right)^{nt}$

   $= 5000\left(1+\dfrac{0.04}{4}\right)^{4(4)}$

   $= 5000(1.01)^{16}$

   ≈ 5862.89

   The account will be worth approximately $5862.89.

16. Approximately $9571.31

17. Approximately $14,511

18. Approximately $23,332

Section 10.3

1. $\log_{10} 1000 = 3$ because $10^3 = 1000$.

2. 3

3. $\log_3 \dfrac{1}{27} = -3$ because $3^{-3} = \dfrac{1}{27}$.

4. -2

5. $\log_{49} 7 = \dfrac{1}{2}$ because $49^{1/2} = 7$.

6. $\dfrac{3}{2}$

7. $\log_7 49 = x$ means $7^x = 49$, thus

   $7^x = 49$

   $7^x = 7^2$

   $x = 2$

   The solution set is {2}.

8. $\{-3\}$

9. $\log_{16} x = \dfrac{1}{2}$ means $16^{1/2} = x$, thus $x = 4$.

   The solution set is {4}.

10. $\left\{\dfrac{1}{10}\right\}$

11. $\log_x 32 = 5$ means $x^5 = 32$, thus

   $x^5 = 32$

   $x^5 = 2^5$

   $x = 2$

   The solution set is {2}.

12. $\left\{\dfrac{1}{2}\right\}$

13. $\log_{11} 11^3 = 3$

14. -4

15. $\log_3 1 = 0$

16. 2

17.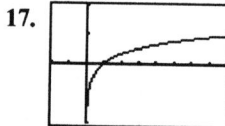

18.

19.

20.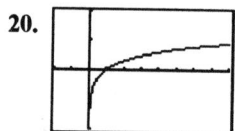

Section 10.4

1. $\log_2 12 + \log_2 6 = \log_2(12 \cdot 6) = \log_2 72$

2. $\log_5(2x - 4)$

3. $\log_3 y + \log_3 5 + \log_3(y - 1) = \log_3 5y + \log_3(y - 1)$
   $= \log_3 5y(y - 1)$

4. $\log_4(7y^2 - 7y - 42)$

5. $\log_{10} 36 - \log_{10} 9 = \log_{10} \dfrac{36}{9} = \log_{10} 4$

6. $\log_5 45$

7. $\log_8 20 + \log_8 3 - \log_8 4 = \log_8 60 - \log_8 4$
   $= \log_8 \dfrac{60}{4}$
   $= \log_8 15$

8. $\log_3(x + 3)$

9. $3\log_3 y = \log_3 y^3$

10. $\log_{10} \dfrac{1}{81}$

11. $-\log_4 x = \log_4 x^{-1} = \log_4 \dfrac{1}{x}$

12. $\log_9 6$

13. $\dfrac{1}{3}\log_7 27 + \log_7 x = \log_7 27^{1/3} + \log_7 x$

$$= \log_7 3 + \log_7 x$$
$$= \log_7 3x$$

14. $\log_4 \dfrac{2}{5}$

15. $3\log_2 x + \log_2 6 - 2\log_2 x = \log_2 x^3 + \log_2 6 - 2\log_2 x$

$$= \log_2 6x^3 - 2\log_2 x$$
$$= \log_2 6x^3 - \log_2 x^2$$
$$= \log_2 \dfrac{6x^3}{x^2}$$
$$= \log_2 6x$$

16. $\log_5 \dfrac{2x^4}{z^3 y^5}$

17. $\log_{10} \dfrac{8}{x} = \log_{10} 8 - \log_{10} x$

18. $\log_5 2 + 3\log_5 y$

19. $\log_3 \dfrac{x^5}{y} = \log_3 x^5 - \log_3 y$

$$= 5\log_3 x - \log_3 y$$

20. $2\log_6 a - 2\log_6 b$

Section 10.5

1. $\log 12 = 1.079181246 \approx 1.0792$

2. 0.7830

3. $\ln 10 = 2.302585093 \approx 2.3026$

4. 1.5581

5. $\log 100{,}000 = \log 10^5 = 5$

6. -2

7. $\ln e^{-5} = -5$

8. $\dfrac{1}{2}$

9. $\log x = -1.15$

$$x = 10^{-1.15} \approx 0.0708$$

10. $e^{3.6} \approx 36.5982$

11. $\ln 2x = 7.1$

$$2x = e^{7.1}$$
$$x = \dfrac{e^{7.1}}{2} \approx 605.9835$$

12. $2\cdot 10^{1.89} \approx 155.2494$

13. $\log(3x+5) = 2.34$

$$3x+5 = 10^{2.34}$$
$$3x = 10^{2.34} - 5$$
$$x = \dfrac{10^{2.34} - 5}{3} \approx 71.2587$$

14. $10^{-1.24} + 6 \approx 6.0575$

15. $\ln(5x-12) = 0.81$

$$5x-12 = e^{0.81}$$
$$5x = e^{0.81} + 12$$
$$x = \dfrac{e^{0.81} + 12}{5} \approx 2.8496$$

16. $\ln(8-3x) = 3.62$

$$8-3x = e^{3.62}$$
$$8 - e^{3.62} = 3x$$
$$\dfrac{8 - e^{3.62}}{3} \approx -9.7792$$

17. $\log_2 9 = \dfrac{\log_{10} 9}{\log_{10} 2} \approx \dfrac{0.9542425094}{0.3010299957} \approx 3.1699$

18. 1.4248

19. $\log_{1/4} 10 = \dfrac{\log_{10} 10}{\log_{10} \frac{1}{4}} \approx \dfrac{1}{-0.6020599913} \approx -1.6610$

20. -1.2920

21. $R = \log\left(\dfrac{a}{T}\right) + B$

$$R = \log\left(\dfrac{300}{2.1}\right) + 4.2 \approx 6.4$$

The earthquake had a magnitude of 6.4 on the Richter scale.

22. 4.4 on the Richter scale

23.

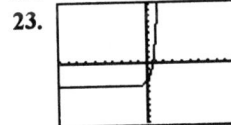

24.

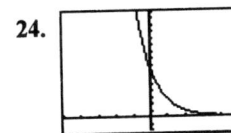

Section 10.6

1. $5^x = 60$

$$\log 5^x = \log 60$$

$$x \log 5 = \log 60$$

$$x = \frac{\log 60}{\log 5}$$

The solution set is $\left\{ \dfrac{\log 60}{\log 5} \right\}$ or $\{2.5440\}$.

2. $\{\ln 15\}$ or $\{2.7081\}$

3. $e^{4x} = 16$

$$\ln e^{4x} = \ln 16$$

$$4x = \ln 16$$

$$x = \frac{\ln 16}{4}$$

The solution set is $\left\{ \dfrac{\ln 16}{4} \right\}$ or $\{0.6931\}$.

4. $\left\{ \dfrac{\log 9}{3 \log 4} \right\}$ or $\{0.5283\}$

5. $2^{3x-4} = 10$

$$\log 2^{3x-4} = \log 10$$

$$(3x - 4) \log 2 = \log 10$$

$$3x - 4 = \frac{\log 10}{\log 2}$$

$$3x = \frac{\log 10}{\log 2} + 4$$

$$x = \frac{\dfrac{\log 10}{\log 2} + 4}{3}$$

The solution set is $\left\{ \dfrac{\dfrac{\log 10}{\log 2} + 4}{3} \right\}$ or $\{2.4406\}$.

6. $\left\{ \dfrac{\ln 24 - 1}{5} \right\}$ or $\{0.4356\}$

7. $\log_5 6 + \log_5 x = 0$

$$\log_5 6x = 0$$

$$6x = 5^0$$

$$6x = 1$$

$$x = \frac{1}{6}$$

The solution set is $\left\{ \dfrac{1}{6} \right\}$.

8. $\left\{ \dfrac{1}{2} \right\}$

9. $\log_4 8 - 3 \log_4 x = 3$

$$\log_4 8 - \log_4 x^3 = 3$$

$$\log_4 \frac{8}{x^3} = 3$$

$$\frac{8}{x^3} = 4^3$$

$$\frac{8}{64} = x^3$$

$$\frac{1}{8} = x^3$$

$$\frac{1}{2} = x$$

The solution set is $\left\{ \dfrac{1}{2} \right\}$.

10. $\{2\}$

11. $\log_7 (4x + 5) = 2$

$$4x + 5 = 7^2$$

$$4x + 5 = 49$$

$$4x = 44$$

$$x = 11$$

The solution set is $\{11\}$.

12. $\{729\}$

13. $\log_2 (x^2 - 11x + 60) = 5$

$$x^2 - 11x + 60 = 2^5$$

$$x^2 - 11x + 60 = 32$$

$$x^2 - 11x + 28 = 0$$

$$(x - 4)(x - 7) = 0$$

$$x = 4, 7$$

The solution set is $\{4, 7\}$.

14. $\{5\}$

15. $y = y_0 e^{0.028t}$

$$y = 26{,}587{,}000 e^{0.028(7)}$$

$$y \approx 32{,}343{,}801$$

There will be approximately 32,343,801 people.

16. 60,883,584 people

17.
$$A = P\left(1+\frac{r}{n}\right)^{nt}$$

$$6000 = 3000\left(1+\frac{0.03}{4}\right)^{4t}$$

$$2 = (1.0075)^{4t}$$

$$\log 2 = \log(1.0075)^{4t}$$

$$\log 2 = 4t(\log 1.0075)$$

$$\frac{\log 2}{4\log 1.0075} = t$$

$$23.19144152 \approx t$$

It takes just over 23 years for the money to double.

18. Approximately 3.3 years.

19.
$$w = 0.00185h^{2.67}$$

$$95 = 0.00185h^{2.67}$$

$$\frac{95}{0.00185} = h^{2.67}$$

$$\left(\frac{95}{0.00185}\right)^{1/2.67} = h$$

$$h \approx 58.1$$

The boy is approximately 58.1 inches tall.

20. Approximately 10.2 pounds per square inch

21. $\{1.11\}$

22. $\{0.27, 2.63\}$

Chapter 10 Practice Test

1. 1

2. $2x^2 + 1$

3. $4x^2 + 12x + 8$

4. $f(x) = \frac{1}{3}x + 8$ and $f^{-1}(x) = 3x - 24$

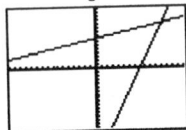

5. Not one-to-one

6. One-to-one

7. Not one-to-one

8. One-to-one
$$f^{-1} = \{(-3, 10), (2, 11), (-1, 12), (-4, 13)\}$$

9. One-to-one
$$f^{-1} = \{(7, \text{Maryland}), (28, \text{Texas}), (17, \text{Ohio}), (36, \text{Nevada}), (18, \text{Louisiana})\}$$

10. $\log_5 9$

11. $\log_3 \dfrac{x^3}{2x-1}$

12. $\log_2 10 + 5\log_2 x + \log_2 y$

13. 3.92

14. 2.4650

15. $\{1\}$

16. $\left\{\dfrac{1}{2}\left(\dfrac{\log 30}{\log 8} + 1\right)\right\}$ or $\{1.3178\}$

17. $\{512\}$

18. $\left\{\dfrac{2}{3}\right\}$

19. $\{14\}$

20. $\left\{\dfrac{27}{8}\right\}$

21. $\{3\}$

22. $\{206.2144\}$

23.

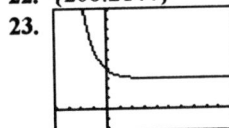

24.

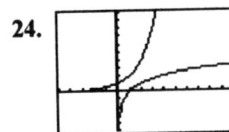

25. $22,408.89

26. Approximately 6.8 years

27. 284,472 people

28. Approximately 8 years

29. $\{0.75\}$

Chapter 11

Section 11.1

1. $a_n = n + 10$

$a_1 = 1 + 10 = 11$	$a_3 = 3 + 10 = 13$
$a_2 = 2 + 10 = 12$	$a_4 = 4 + 10 = 14$

2. $-\dfrac{1}{2}, -1, -\dfrac{3}{2}, -2$

3. $a_n = 3n - 1$

$a_1 = 3(1) - 1 = 2$	$a_3 = 3(3) - 1 = 8$
$a_2 = 3(2) - 1 = 5$	$a_4 = 3(4) - 1 = 11$

4. $\dfrac{2}{1}, \dfrac{2}{2}, \dfrac{2}{3}, \dfrac{2}{4}$

5. $a_n = (-1)^n (n+2)$

$a_{11} = (-1)^1 (1+2) = -3$

$a_{12} = (-1)^2 (2+2) = 4$

$a_{13} = (-1)^3 (3+2) = -5$

$a_{14} = (-1)^4 (4+2) = 6$

6. $\dfrac{3}{1}, \dfrac{4}{2}, \dfrac{5}{3}, \dfrac{6}{4}$

7. $a_n = 3^n$

$a_1 = 3^1 = 3 \qquad a_3 = 3^3 = 27$

$a_2 = 3^2 = 9 \qquad a_4 = 3^4 = 81$

8. $9, 6, 1, -6$

9. $a_n = \dfrac{n-2}{n+4}; \; a_{18}$

$a_{18} = \dfrac{18-2}{18+4} = \dfrac{16}{22}$

10. 64

11. $a_n = (-1)^n (3n-2); \; a_{11}$

$a_{11} = (-1)^{11} (3 \cdot 11 - 2) = -31$

12. -350

13. $6, 7, 8, 9, \ldots$

Each term is 1 more than the previous term, and each term is 5 more than its corresponding number in the sequence of natural numbers. A general term might be $a_n = n+5$.

14. $a_n = \dfrac{1}{n^2}$

15. $7, 10, 13, 16, \ldots$

Each term is 4 more than three times its corresponding number in the sequence of natural numbers. A general term might be $a_n = 3n+4$.

16. $(-2)^{n+1}$

17. 200, 280, 360, 440, 520, 600 customers

18. $a_n = 100{,}000 - 12{,}500n$; $0

19. $a_n = \sqrt{2n-1}$

1, 1.7321, 2.2361, 2.6458

20. 0, 0.3466, 0.3662, 0.3466

Section 11.2

1. $a_1 = -16; d = 5 \Rightarrow a_n = -16 + (n-1)5$

$a_1 = -16$

$a_2 = -16 + (2-1)5 = -11$

$a_3 = -16 + (3-1)5 = -6$

$a_4 = -16 + (4-1)5 = -1$

2. 30, 22, 14, 6

3. $a_1 = 10; r = 4 \Rightarrow a_n = 10(4)^{n-1}$

$a_1 = 10$

$a_2 = 10(4) = 40$

$a_3 = 10(4)^{3-1} = 160$

$a_4 = 10(4)^{4-1} = 640$

4. $-4, 2, -1, \dfrac{1}{2}$

5. $a_n = a_1 + (n-1)d$

$a_n = 5 + (n-1)12$

$a_{10} = 5 + (10-1)12 = 5 + 9 \cdot 12 = 5 + 108 = 113$

6. -13

7. $r = \dfrac{200}{1000} = \dfrac{1}{5}$

$a_n = 1000 \left(\dfrac{1}{5} \right)^{n-1}$

$a_5 = 1000 \left(\dfrac{1}{5} \right)^{5-1} = \dfrac{8}{5}$

8. $-300{,}000$

9. $r = \dfrac{12}{3} = 4$

$a_2 = a_1 (4)^1$

$3 = a_1 (4) \Rightarrow a_1 = \dfrac{3}{4}$

10. 26.25

11. 3, 15, 75

$r = \dfrac{15}{3} = 5$

12. $d = 2$

13. 6, 26, 46

$d = 26 - 6 = 20$

14. $r = -2$

15. $a_n = 26{,}000(1.04)^{n-1}$

$a_6 = 26{,}000(1.04)^{6-1} \approx \$31{,}632.98$

16. \$31,500

17. $a_1 = 8{,}000{,}000 \quad a_2 = 13{,}000{,}000 \quad a_3 = 18{,}000{,}000$

$d = 13{,}000{,}000 - 8{,}000{,}000 = 5{,}000{,}000$

$a_n = 8{,}000{,}000 + (n-1)5{,}000{,}000$

$2000 \Rightarrow n = 6$

$a_6 = 8{,}000{,}000 + (6-1)5{,}000{,}000 = 33{,}000{,}000$

18. Approximately \$660 million

Approximately \$1462 million

Section 11.3

1.
$$\sum_{i=1}^{6}(i-5) = -4+(-3)+(-2)+(-1)+0+1 = -9$$

2. 50

3.
$$\sum_{i=1}^{5}\frac{i}{5} = \frac{1}{5}+\frac{2}{5}+\frac{3}{5}+\frac{4}{5}+\frac{5}{5} = 3$$

4. 12

5.
$$\sum_{i=2}^{5}i(i+3) = 2(2+3)+3(3+3)+4(4+3)+5(5+3)$$
$$= 2\cdot5+3\cdot6+4\cdot7+5\cdot8$$
$$= 96$$

6. 120

7. $3 + 15 + 75 + 375$
$$r = \frac{15}{3} = 5 \qquad a_n = 3(5)^{n-1}$$
$$\sum_{i=1}^{4}3(5)^{i-1}$$

8.
$$\sum_{i=1}^{5}(-9+2i)$$

9. $6 + 26 + 46 + 66$
$$d = 26-6 = 20, a_n = 6+(n-1)20 = 20n-14$$
$$\sum_{i=1}^{4}(20i-14)$$

10.
$$\sum_{i=1}^{3}10(-2)^{i-1}$$

11. $5 + 10 + 20 + 40 + 80$
$$r = \frac{10}{5} = 2 \qquad a_n = 5(2)^{n-1}$$
$$\sum_{i=1}^{5}5(2)^{i-1}$$

12.
$$\sum_{i=1}^{7}2i$$

13.
$$\sum_{i=1}^{4}2i^2 = 2(1)^2 + 2(2)^2 + 2(3)^2 + 2(4)^2$$
$$= 2+8+18+32 = 60$$

14. 4833

15. $14, 13, 12, \ldots, 2, 1$; 105 quilt blocks

16. $110,408.06

Section 11.4

1. 10, 15, 20, 25, 30, 35, 40, 45, 50, 55
$$S_n = \frac{n}{2}(a_1+a_n) = \frac{10}{2}(10+55) = 5(65) = 325$$

2. 2728

3. $-16, -8, -4, \ldots$
$$S_n = \frac{a_1(1-r^n)}{1-r} = \frac{-16[1-(.5)^6]}{1-0.5} = -31.5$$

4. -30

5.
$$\frac{1}{2}, \frac{3}{4}, 1, \frac{5}{4}, \frac{3}{2}, \frac{7}{4}, 2, \frac{9}{4}$$
$$S_n = \frac{n}{2}(a_1+a_n) = \frac{8}{2}\left(\frac{1}{2}+\frac{9}{4}\right) = 4\left(\frac{11}{4}\right) = 11$$

6. 111.1111

7.
$$S_n = \frac{n}{2}(a_1+a_n) = \frac{25}{2}(1+25) = \frac{25}{2}(26) = 325$$

8. 5050

9. $1, \frac{7}{8}, \frac{49}{64}, \ldots \qquad r = \frac{7}{8}$
$$S_\infty = \frac{a_1}{1-r} = \frac{1}{1-\frac{7}{8}} = 8$$

10. 343

11. $512, 64, 8, \ldots$
$$S_\infty = \frac{a_1}{1-r} = \frac{512}{1-\frac{1}{8}} = \frac{4096}{7}$$

12. $\frac{20}{3}$

13. $-7, -\frac{7}{3}, -\frac{7}{9}, \ldots$
$$S_\infty = \frac{a_1}{1-r} = \frac{-7}{1-\frac{1}{3}} = -\frac{21}{2}$$

14. -25

15. $4000, 2000, 1000, \ldots$
$$S_n = \frac{a_1(1-r^n)}{1-r} = \frac{4000[1-(.5)^{10}]}{1-0.5} \approx 7992$$

Section 11.5

1. $(z-y)^4 = z^4 - 4z^3y + 6z^2y^2 - 4zy^3 + y^4$

2. $(s+t)^8 = s^8 + 8s^7t + 28s^6t^2 + 56s^5t^3 + 70s^4t^4$
$$+ 56s^3t^5 + 28s^2t^6 + 8st^7 + t^8$$

3. $\frac{4!}{7!} = \frac{4!}{7\cdot6\cdot5\cdot4!} = \frac{1}{210}$

4. 220

5.
$$(w+z)^4 = w^4 + \frac{4}{1!}w^3z + \frac{4\cdot3}{2!}w^2z^2 + \frac{4\cdot3\cdot2}{3!}wz^3 + z^4$$
$$= w^4 + 4w^3z + 6w^2z^2 + 4wz^3 + z^4$$

6. $m^6 - 6m^5n + 15m^4n^2 - 20m^3n^3 + 15m^2n^4 - 6mn^5 + n^6$

7.
$$(a-3b)^4 = a^4 + \frac{4}{1!}a^3(-3b) + \frac{4\cdot3}{2!}a^2(-3b)^2$$
$$+ \frac{4\cdot3\cdot2}{3!}a(-3b)^3 + \frac{4\cdot3\cdot2\cdot1}{4!}(-3b)^4$$
$$= a^4 - 12a^3b + 54a^2b^2 - 108ab^3 + 81b^4$$

8. $32x^5 + 80x^4y + 80x^3y^2 + 40x^2y^3 + 10xy^4 + y^5$

9.
$$(2q+3r)^3 = (2q)^3 + \frac{3}{1!}(2q)^2(3r) + \frac{3\cdot2}{2!}(2q)(3r)^2$$
$$+ \frac{3\cdot2\cdot1}{3!}(3r)^3$$
$$= 8q^3 + 36q^2r + 54qr^2 + 27r^3$$

10. $625 - 1000p + 600p^2 - 160p^3 + 16p^4$

11. $(q+2r)^{11}, r+1 = 7$
$$\frac{n!}{r!(n-r)!}a^{n-r}b^r = \frac{11!}{6!\,5!}q^{11-6}(2r)^6$$
$$= 462q^5(64r^6)$$
$$= 29{,}568q^5r^6$$

12. $1215x^4y^2$

13. $(w-7)^8, r+1 = 5$
$$\frac{n!}{r!(n-r)!}a^{n-r}b^r = \frac{8!}{4!\,4!}w^{8-4}(-7)^4$$
$$= 70w^4(2401)$$
$$= 168{,}070w^4$$

14. b^9

Chapter 11 Practice Test

1. $1, \dfrac{4}{3}, \dfrac{9}{5}, \dfrac{16}{7}$

2. 0, 4, 10, 18

3. 316

4. 14,161

5. $a_n = \dfrac{6}{7}(2)^{n-1}$

6. $a_n = 40 + (n-1)(-16)$

7. 776

8. 968

9. 18

10. $\dfrac{10}{3}$

11. 35

12. 460

13. $8a^3 - 12a^2b + 6ab^2 - b^3$

14. $y^5 + 15y^4 + 90y^3 + 270y^2 + 405y + 243$

15. $y^4 + 12y^3r + 54y^2r^2 + 108yr^3 + 81r^4$

16. $x^6 - 6x^5y + 15x^4y^2 - 20x^3y^3 + 15x^2y^4 - 6xy^5 + y^6$

17. 900 people

18. 240 shrubs

Practice Final Exam #1

1. True

2. False

3. −262

4. −19

5. 3

6. 8

7. $\dfrac{9}{2}$

8. $2(q-6) < 10$

9. Distributive property

10. $5n + 25q$

11. $\{-6\}$

12. $\{200\}$

13. $\left\{\dfrac{4}{3}, 4\right\}$

14. $y = \dfrac{9x-11}{32}$

15. $\dfrac{V}{lw} = h$

16. $[65, \infty)$

17. $\left[\dfrac{9}{5}, \dfrac{16}{5}\right)$

18. $\left(-\infty, -\dfrac{7}{2}\right) \cup \left(\dfrac{21}{2}, \infty\right)$

19. 136.8

20. $12,739.96

21. Quadrant II, Quadrant III, Quadrant I, Quadrant IV

22.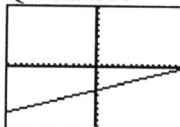

5

23.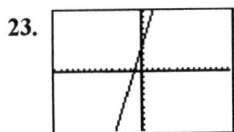

24.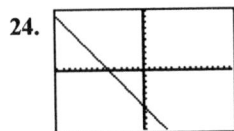

25.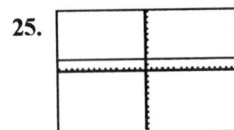

26. $x = 8$

27. $y = -6$

28. $y = -2x + 11$

29. $y = -2x + 16$

30. Domain: {0, 3, –2, 4}; Range: {9, 2, 5, –4}; No

31. –10

32. –22

33. {(3, 0)}

34. $\left\{\left(5, 2, -\dfrac{5}{3}\right)\right\}$

35. {(0, –1)}

36. {(2, –1)}

37. {(6, 0)}

38. {(1, –2, 2)}

39. 7.5 pounds of Brazilian Rain Forest Coffee, 22.5 pounds of Blue Mountain Dark Roast Coffee

40. 620 rooms on the weekend, 1400 rooms during the week

41. $\dfrac{1}{32x^5 y^5}$

42. $\dfrac{m^7}{4n^3}$

43. 4.72×10^4

44. 0.00671

45. $3x^2 - 11x + 12$

46. $25x^2 - 64$

47. $3z(3z^2 - z - 7)$

48. $(x + 7)(x + 4)$

49. {–3, 3}

50.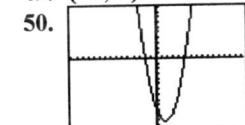

51. $\left\{x \mid x \text{ is a real number and } x \neq \dfrac{20}{3}\right\}$

52. $\dfrac{(2 - x)}{(9x - 15)}$

53. $\dfrac{y}{8(y - 5)}$

54. $\dfrac{11x - 6}{(x + 6)(x + 2)}$

55. $\dfrac{38}{3}$

56. $\dfrac{5ab}{c} + \dfrac{2a}{c} - \dfrac{9c}{a}$

57. $3x^3 + 10x^2 + 57x + 205 + \dfrac{625}{x - 5}$

58. {30}

59. $\left\{\dfrac{30}{7}\right\}$

60. 75

61. $-x^{16}$

62. 8

63. $\dfrac{8\sqrt{5}}{5y}$

64. $y\sqrt{5y}$

65. $\sqrt{6} - 6\sqrt{2} + 8\sqrt{3} - 48$

66. {2, 7}

67. $6 + 3i$

68. 68

69. $\left\{-4 - 4\sqrt{2}, -4 + 4\sqrt{2}\right\}$

70. $\left\{-10 - \sqrt{94}, -10 + \sqrt{94}\right\}$

71. {–7, 2}

72. $\left\{\sqrt{2}, -\sqrt{2}, 3i, -3i\right\}$

73. {–1, 1}

74. (–2, 5)

75.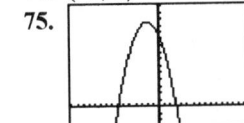
Vertex: (–2, 25)

76. $\dfrac{7 + 7\sqrt{7}}{2}$ feet; 12.8 feet

77. $d = 4\sqrt{10}$; (1, 2)

78. $d = 2\sqrt{29}$; (4, –2)

79.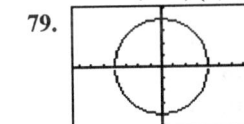

80.

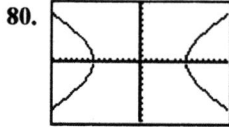

81.

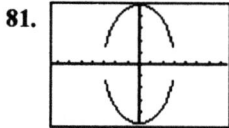

82.

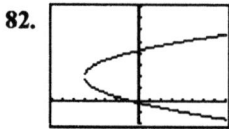

83. $\{(-4.5, 6), (0, 6)\}$

84.

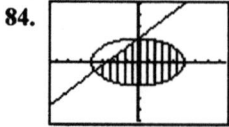

85. (a) 2; (b) $6x^2 - 1$

86. $\dfrac{x - 11}{-2}$

87. Yes

88. $\log_3 (4x^3 + 3x^2)$

89. $\log_5 8 + 4\log_5 x + 2\log_5 z$

90. $\{3\}$

91. $\left\{\dfrac{4}{3}\right\}$

92. $\{-1.30\}$

93. $a_n = \dfrac{n^3}{2n^2 + 5}$

$\dfrac{1}{7}, \dfrac{8}{13}, \dfrac{27}{23}, \dfrac{64}{37}$

94. 1255

95. $a_n = 100 - 33(n - 1)$

96. 420

97. 48

98. 120

99. $r^5 - 5r^4 s + 10r^3 s^2 - 10r^2 s^3 + 5rs^4 - s^5$

100. 119 shrubs

Practice Final Exam #2

1. b	26. c	51. c	76. c
2. a	27. b	52. b	77. a
3. c	28. d	53. a	78. d
4. a	29. d	54. b	79. b
5. c	30. b	55. c	80. d
6. d	31. a	56. d	81. c
7. b	32. c	57. d	82. a
8. b	33. a	58. b	83. c
9. a	34. b	59. c	84. c
10. c	35. c	60. a	85. b
11. a	36. c	61. d	86. a
12. c	37. a	62. b	87. a
13. d	38. a	63. b	88. b
14. b	39. d	64. b	89. d
15. c	40. c	65. a	90. b
16. c	41. b	66. c	91. b
17. a	42. e	67. d	92. c
18. d	43. d	68. c	93. a
19. a	44. b	69. b	94. b
20. c	45. c	70. a	95. c
21. c	46. b	71. d	96. d
22. b	47. a	72. b	97. a
23. d	48. d	73. c	98. b
24. a	49. c	74. b	99. a
25. b	50. b	75. a	100. c